I0867734

GARDEN CULTURE
of the
TWENTIETH CENTURY

ex horto

DUMBARTON OAKS TEXTS
IN GARDEN AND LANDSCAPE STUDIES

LEBERECHT MIGGE
DIE GARTENKULTUR DES 20. JAHRHUNDERTS

GARDEN CULTURE *of the* TWENTIETH CENTURY

LEBERECHT MIGGE,
author

DAVID H. HANEY,
editor and translator

DUMBARTON OAKS RESEARCH LIBRARY AND COLLECTION
WASHINGTON, D.C.

© 2013 Dumbarton Oaks
Trustees for Harvard University, Washington, D.C.
All rights reserved.
Printed in the United States of America by Sheridan Books, Inc.

Library of Congress Cataloging-in-Publication Data
Migge, Leberecht, 1881–1935.
[Gartenkultur des 20. Jahrhunderts. English]
Garden culture of the twentieth century / Leberecht Migge, author ; David H. Haney, editor and translator. — 1st English ed.
p. cm. — (Ex Horto: Dumbarton Oaks texts in garden and landscape studies)
Translation of: Die Gartenkultur des 20. Jahrhunderts, originally published in 1913.
ISBN 978-0-88402-388-3 (pbk. : alk. paper)
1. Landscape gardening—Germany. 2. Gardens—Germany.
I. Haney, David H. II. Dumbarton Oaks. III. Title.
IV. Series: Ex Horto: Dumbarton Oaks texts in garden and landscape studies.
SB469.35.G3M5413 2013
635.90943—dc23
012046891

Series editor: John Beardsley
Managing editor: Sara Taylor
Series design: Melissa Tandysh

Frontispiece: Original cover stamp of *Die Gartenkultur des 20. Jahrhunderts* (Jena, Germany: Eugen Diederichs, 1913).

www.doaks.org/publications

CONTENTS

FOREWORD TO THE ENGLISH TRANSLATION

This volume makes available in English for the first time the full text of one of the least known but most intriguing books in the history of modern landscape architecture, Leberecht Migge's *Die Gartenkultur des 20. Jahrhunderts* (*Garden Culture of the Twentieth Century*, hereafter, *Garden Culture*). Translated by David H. Haney, the author of the first study of Migge's life and work in English, this edition appears on the centenary of the book's original publication in 1913.

Leberecht Migge was among the most innovative landscape architects of the early twentieth century, and he focused especially on the environmental challenges of the industrial metropolis. With work ranging from large urban parks to housing settlements with allotment gardens, he sought to create functional green spaces that would both accommodate and improve the social conditions of modern life. In the years before World War I, he began to develop a design language that would express these ideals, a spare and geometrical approach that broke with the prevailing practices of landscape gardening in late nineteenth-century Germany.

Migge's social and political engagements were already evident in *Garden Culture*, his first major text. This work, which continued to be a touchstone for the theoretical debates of the Weimar era, argued that landscape gardening was an outworn approach to environmental design based on a bourgeois, aestheticized relation to nature. Migge maintained that it was only by analyzing and updating the traditional forms of productive landscapes—cultivated fields and vegetable plots—that modern society could create habitable cities. *Garden Culture* was richly illustrated with drawings and photographs of such landscapes, together with modern examples that included some of his own designs. These functioned as a visual argument running parallel with the text; they have been retained in full in this edition. Although the book was not translated into English, it was nevertheless influential in the Anglo-American world in the years following its publication. This stemmed in part from Migge's inclusion of several American parks as positive examples of green spaces used for sports and other activities. Within professional circles, the book was best known through Henry Hubbard and Theodora Kimball's discussion of it in *An Introduction to the Study of Landscape Design* (1917) and from Kimball's review in the October 1914 issue of *Landscape Architecture* magazine. Migge received some renewed attention in the 1960s, and he is today increasingly recognized for anticipating the contemporary emphasis on "green" or sustainable architecture, landscape design, and urban planning. His book is important not only as a historical document, revealing the theoretical underpinnings of the modernist approach to landscape design, but also as a statement of functionalist principles, with their social, economic, and ethical implications.

As David Haney makes clear in a detailed introduction to this volume, Migge's work was rooted in the German Werkbund movement. Founded in 1906, the movement was shaped by Nietzsche's aesthetics, especially the conviction that art is not a vehicle for disinterested aesthetic contemplation but rather is part of life. The Werkbund advanced cultural reform and social and spiritual renewal through the improvement of production and consumption standards in the applied arts, design, and manufacturing. Migge was closely associated with Hermann Muthesius, one of the founding figures of the Werkbund, collaborating with him for several years. Migge would translate the Werkbund's principles to landscape, transforming the understanding of gardens from places of passive aesthetic contemplation to spaces of production and active recreation. Building on the Werkbund's commitment to reform of the domestic realm, he emphasized utility, organization, and planning over aesthetic effects and developed a notion of "rhythm" as a defining concept that brings together all the elements of a composition. Plants were to be used to create structure and define space, as opposed to serving visual or scenic ends; to achieve these results, Migge paid special attention to their color and form, in accordance with the "honest" use of materials advocated in the philosophy of the Werkbund.

With Migge, however, these concepts left the arena of theoretical debate and acquired a social dimension: his garden culture was specifically designed to suit modern metropolitan conditions. His important public projects, such as the Mariannenpark in a suburb of Leipzig and the city park for the new town of Rüstringen (now Wilhelmshaven), were meant to encourage an active outdoor lifestyle with an emphasis on hygiene and recreation, and were designed with the human body and its movement in mind. Migge approached garden design on an urban scale, proposing a typology of landscapes based on ownership, function, and location and related to contemporaneous notions of the "garden city" and metropolitan planning. Migge was especially aware of the creation of extensive urban parks in the United States, including Charles Eliot's plans for the Boston metropolitan park system. In this way, he contributed to the development of international support for the distribution of green spaces within the urban fabric.

It is both for Migge's historical understandings of the city as a productive landscape and his continued relevance to ideas about sustainable design that

Dumbarton Oaks offers this translation of his most important book. This volume is one of two titles being published simultaneously by Dumbarton Oaks in a new series of translations of significant texts in the histories of garden design and landscape architecture. The other is a full transcription and translation of a very unusual handwritten manuscript in the Rare Book Collection at Dumbarton Oaks by the nineteenth-century German court gardener Hans Jancke: a journal of Jancke's extended stay at Knowsley, the seat of the Earls of Derby near Liverpool, in 1874–75. Additional translations from other languages will appear in the coming years.

JOHN BEARDSLEY
Director, Garden and Landscape Studies
Dumbarton Oaks

TRANSLATOR'S PREFACE AND ACKNOWLEDGMENTS

I have wanted to translate *Garden Culture* into English since I first read it in 1999. That was the year before I moved to Berlin in 2000 to begin a five-year PhD dissertation research project, later published as *When Modern Was Green: Life and Work of Landscape Architect Leberecht Migge* (2010). Among all of Migge's publications, I believe that *Garden Culture* is the most sophisticated and significant, for reasons that I have tried to elaborate in the Introduction. I am thus very grateful to Dumbarton Oaks for inviting me to undertake this translation. It also seems highly appropriate that this translation is appearing in time for the hundredth anniversary of the volume's original publication.

In the process of translating *Garden Culture* into English, I have shown this book to numerous knowledgeable people, including native German-speaking landscape historians. All of them have noted just how eccentric and convoluted Migge's writing style is. In addition, he used antiquated terminology, coined his own words, and was extremely sarcastic. As a translator, I have tried to maintain as much of the style of the original as possible, including the ever-present sarcasms, while maintaining an overall clarity. But a few sentences were so elliptical and convoluted that I felt it necessary to include an explanatory note.

Although a number of people have assisted me with the translation process, the following three were the most instrumental: Professor Sonja Dümpelmann (Department of Landscape Architecture, Harvard University), Dr. Clemens Alexander Wimmer (Director, The German Horticultural Library, Berlin Technical University), and Dr. Pamela Selwyn (professional translator, Berlin). Without their help, I could not have brought this project to completion, although as the translator, I assume full responsibility for the final product.

My deepest gratitude goes to my employer, the University of Kent, who graciously granted me an early sabbatical semester, without which it would have been impossible to undertake this project. Among the people at Kent whom I wish to thank are Professor Karl Leydecker (Dean of Humanities), Professor Laurence Goldstein (Acting Dean, autumn term 2011), Professor Don Gray (Head of

School, Architecture), and Dr. Timothy Brittain-Catlin (Director of Research, Architecture). My sabbatical was granted in conjunction with my activities in the Centre for Research in European Architecture (CREAte, Kent, Professor Gordana Fontana-Giusti, Director). I would also like to thank Dumbarton Oaks for funding a one-month stay in Washington, D.C., in April 2012, for the purpose of collaborating with their staff. All of the other research fellows there at the time were indeed very supportive, but I would particularly like to thank Dr. Michael Herchenbach for his help with German terms and Dr. Robin Veder (Pennsylvania State University) for our discussions on "rhythm" and German analytical psychology. Of course, I am indebted to all of the staff in Garden and Landscape Studies: the director, Professor John Beardsley, and his assistant, Jane Padelford, as well as Michael Lee, then postdoctoral associate. Among them, I would especially like to express my gratitude to Professor Lee (now at the University of Virginia) for his continued support and enthusiasm for this project, and for conceiving of it in the first place. Last but not least, I would like to acknowledge the Dumbarton Oaks publications team (Sara Taylor and Kathleen Sparkes) for their efforts and creative input as to the design of the book.

And how could I conclude this list of acknowledgments without mentioning the long-deceased, extremely eccentric but never dull Leberecht Migge, whose legacy has so enriched my life over the past several years.

DAVID H. HANEY
Berlin, July 2012

figure 1

Portrait of Leberecht Migge (1881–1935), taken ca. 1919 in Hamburg.

Courtesy of the Barkenhoff-Stiftung Worpswede, Germany, from the Migge Collection.

TRANSLATOR'S INTRODUCTION

At first glance, Leberecht Migge's 1913 book, *Garden Culture of the Twentieth Century*, may appear to be somewhat peculiar for a treatise on gardening. Migge opens with a dramatic call to "sweep aside the great rubbish and detritus that covers over the field of gardening,"[1] and concludes in unmistakably militaristic tones by proclaiming that in the "new" garden culture, "there will be conquerors and conquered."[2] Apparently not the typical rhetoric that one would expect to find with respect to gardening in 1913. Yet if one considers Migge's book within the discourse of the German Werkbund, then it is not so unusual. The German Werkbund was an organization composed of leading designers, manufacturers, and cultural critics, all dedicated to the improvement of culture through the enlightened control of systems of production and consumption. Rather than rejecting capitalism and the industrial metropolis, they hoped to work within these contexts by applying high standards of taste and behavior to all levels of design, as the architect Hermann Muthesius expressed it: "from the sofa cushion to the city plan."[3] Migge himself was a Werkbund member and a working colleague of Muthesius, one of its most active members. Migge undoubtedly conceived of *Garden Culture* as a means of introducing garden design into wider Werkbund discourse. Like other Werkbund members, Migge believed that cultural reform began with the domestic culture of the well-to-do, which it was hoped would influence the whole of society by example. Their program was inherently paternalistic and elitist, even though they wanted to effect change at a "mass" scale. Muthesius was one of the leading proponents of the reform of domestic culture, and although he expressed his belief that the garden and dwelling should function as one unit, as an architect he focused primarily on the house. Migge essentially picked up where Muthesius had left off, concentrating on the garden and systematically analyzing the way it was produced and used.

Other garden writers of the period also framed their discussions in terms of domestic design reform and the "new" garden, for example: in Germany, Fritz Encke, *The House Garden* (1907), Willy Lange, *Garden Design of the New Era* (1907), and Camillo Schneider, *The Landscape Mode of Garden Design* (1907); in England, Reginald Blomfield, *The Formal Garden in England* (1892), Thomas H. Mawson, *The Art and Craft of Garden Making* (1901), and William Robinson, *The Wild Garden* (1870); in France, André Vera, *The New Garden* (1912); and in the United States, Theodora Kimball and Henry Hubbard, *An Introduction to the Study of Landscape Design* (1917).[4] These other garden treatises, however, lack the protracted discussions of the "national economy" and the importance of international competition found here—rhetoric that Migge clearly borrowed from Werkbund analyses of mass production and export. While other German garden treatises may have been nationalistic to various degrees, arguing for a particular form of "German" garden expression, only Migge concluded with a militaristic "call to arms," again an indication of the tone used within Werkbund circles.[5]

Yet there was another critical area influencing Migge's thinking at the time: the newly emerging field of urban planning (*Städtebau*), which had begun to be established as a discipline in Germany in the 1870s. Here Migge was not alone, for internationally a number of other garden designers and writers wished to make a claim for their own field within urban planning. Two important contemporary examples were the English garden designer and planner Thomas Mawson's *Civic Art* (1911) and the Parisian garden designer and planner J. C. N. Forestier's *Great Cities and Park Systems* (1906); the German garden architect Hugo Koch's book, *Garden Design and City Planning*, appeared in 1914, a year after Migge's book.[6] Not insignificantly, Mawson and Forestier discussed domestic gardens and public parks in fully separate publications, in contrast to Migge's more comprehensive study here.

The person who undoubtedly had the greatest impact on Migge's thinking about urban planning was the German economist and planner Werner Hegemann.[7] As an economist concerned with political and social matters, Hegemann saw the public park as a critical instrument for improving urban life for the masses. In 1911, Hegemann published a brochure titled *A Park Book*, showcasing new urban parks in the United States, primarily in Chicago and Boston.[8] The firm that Migge worked for financed the publication, and a section illustrating the firm's recent residential garden designs was appended, but the two sections were not really integrated.[9] In his 1913 *Garden Culture*, Migge obviously strove to synthesize these two bodies of thought within one overall system. Here, the public park is treated with the same design emphasis as the private garden, while the private garden is also presented as an agent of social change on a mass scale. One of Migge's central tactics in *Garden Culture* was to layer the usual rhetoric of one of these areas of discussion over the other. Thus, what Migge succeeded in doing, in contrast to his contemporaries, was to synthesize thinking from garden design, urban planning, and the studies on mass design and trade within the Werkbund. From the latter, particularly from Muthesius, Migge took the concept of the repetitive mass type, which he then applied to all forms of garden and public green space as a means of conceptually organizing the whole of his garden culture.

But in the end, for Migge as a gardener the most obvious stimulus to his thinking came from the field of garden design. In the opening years of the twentieth century, followers of the "garden reform" movement had been arguing that the typical nineteenth-century naturalistic garden was wholly outmoded and should be replaced with a new, more functional design approach based on geometric planning. As a champion of garden reform himself, Migge in *Garden Culture* made the obligatory disparaging references to the old-fashioned landscape garden. Another contemporary thinker whose publications Migge certainly would have been familiar with, the architect, painter, and cultural critic Paul Schultze-Naumburg, aimed his own critique of garden design not at the larger-scale landscape park but at smaller imitations of "nature" in the suburban house garden. But it is important to understand the overall context of the garden reform movement, for the primary driving forces behind this change were as much philosophical as they were practical. For example, Schultze-Naumburg called for the reform not only of the garden but also of clothing, asserting the need for greater bodily movement to enable more active, outdoor living.[10] Reformers such as Migge and Schultze-Naumburg believed that the designs and cultural mores of the late nineteenth century were both artificial and repressive.

figure 2

Illustrations from *Kulturarbeiten* by Paul Schultze-Naumburg, typically showed a positive example (left) and a negative "counter" example (top) of landscapes, in this case of two garden basins. (Schultze-Naumburg nearly always took his own photographs.)

Reproduced from Paul Schultze-Naumburg, *Kulturarbeiten*, vol. 2, *Gärten* (Munich: Callwey, 1912), 226 (left) and 229 (top).

The protagonists of all of these reform-era movements, Migge included, painted a black-and-white picture of the generation preceding them, ignoring any positive or sympathetic qualities in the older design work. Instead, they emphasized the newness and originality of their own proposals, often a somewhat disingenuous tactic. As with all propaganda, some elements in *Garden Culture* are positive and convincing, while others are overly ambitious and idealistic to the point of losing credibility. Yet Migge makes many original, far-reaching observations in relation to his vision of an all-encompassing "garden culture" that may still be considered relevant today.

Biographical Sketch

Migge was the product of his background and training. Born in Gdańsk (Danzig) to a modest merchant family in 1881, he went to work in a commercial nursery in his youth because of his father's early death.[11] In 1899, at the age of eighteen, he began his formal training at a government-run school for gardeners in Oranienburg near Berlin, where he remained for two years. There, he gained his understanding of horticultural practice and garden design; among the garden writers that students were encouraged to read were Humphry Repton and Prince Hermann von Pückler-Muskau. In 1901, young Leberecht began as a designer at the firm of Jakob Ochs in Hamburg, which, as the wealthiest city in Germany, provided a solid base of clients who wanted the latest version of the garden as an expression of their discriminating taste. Migge published his first garden designs in 1907; he became a known figure when he published a brochure in response to the debate on the design competition for the proposed Hamburg City Park in 1909.[12] The ambitious young man introduced himself to key figures of the day, including Hermann Muthesius and the Hamburg art critic Alfred Lichtwark. The ideas of both of these men are evident in *Garden Culture*; indeed, without the influence of Muthesius this book is scarcely conceivable. Shortly after Migge published *Garden Culture* in 1913, he left the Ochs firm because of "irreconcilable differences" and set up his own practice in the wealthy Hamburg suburb of Blankenese, where he was then living.

After 1918, Migge more or less abandoned the residential garden for the well-to-do as a theoretical focus. While he was clearly concerned with the "masses" and small gardens for the working man in *Garden Culture*, these were discussed in general terms as a means of improving living conditions. In his more politically emphatic texts after World War I, he argued that "everyone" could be economically self-sufficient if provided with his own small garden; thus, he devoted the rest of his career to the design of allotment garden colonies and settlements.[13] Throughout his life, he was above all a passionate gardener who, as an emotional and obsessive man, thought that everyone else should be too. This is one reason for his vision in 1913 of "garden culture" as an all-pervasive element in society, acting as a panacea for many ills. But in light of the originality of his ideas and his important contributions to the field at the time, one may forgive him if he was sometimes overzealous and bombastic. Migge died suddenly in May 1935 of kidney failure, at the beginning of the National Socialist period, isolated as much or more because of interpersonal conflicts than because of his political persuasions.

English Critiques of the "Landscape" Garden: Wild Flowers and the Architectural

Migge's use of the landscape garden as a negative foil to his own proposals was typical of garden reform rhetoric during this period. Throughout the nineteenth century, up until the eve of World War I, Germany was highly influenced by ideas and fashions imported from England (although the converse was rarely the case). Both the landscape garden itself and the late nineteenth-century reaction against it originated in England. What German garden writers in the early twentieth century meant by "landscape garden" may be characteristically described by the term "gardenesque," which was coined by the English gardener and writer J. C. Loudon in the 1830s.[14] The gardenesque arose in response to the boredom allegedly experienced with designed picturesque landscapes that were so "natural" that they couldn't be distinguished from the surrounding countryside. Loudon and others believed that art and artistic effects should be introduced to further stimulate the senses and heighten the aesthetic effect. Thus, elaborate compositions of plants were inserted within naturalistic scenes, often furnished with exotic hothouse plants in warmer months. In this way, the picturesque landscape became "garden-like" or garden*esque*. One popular device, dubbed "carpet bedding"—that is, the cutting of elaborately patterned flower beds into lawn areas—was so named because it resembled the lush designs on machine-woven carpets of the period. Such gardening effects were intentionally elaborate and costly, although the overall setting was still intended as a perfected representation of natural scenery.

It is not hard to imagine that this kind of garden perfectly complemented the highly ornate domestic furnishings displayed at the Great Exhibition of the Works of Industry of All Nations, held in London in 1851.[15] And just as the founders of the English Arts and Crafts movement reacted against such elaborate historicist furnishings and architecture, so too they decried the showy planting designs of the gardenesque. Refuge was sought in the old-fashioned cottage garden, both real and imagined, as a return to a less complicated, more honest period before the Industrial Revolution. Simple geometric planting beds for flowers and vegetables were revived, as were old-fashioned plants such as sunflowers. At his Red House, William Morris preserved an old vegetable garden and orchard and maintained a topiary hedge.[16] Perhaps the first serious attack on the gardenesque with any theoretical depth was made by the gardener William Robinson, who gained followers through his book *The Wild Garden*, first published in 1870.[17] Robinson focused attention on the edges and neglected areas of the garden, proposing that hardy wild flowers could be planted there to create quiet, harmonious effects, especially in relation to surrounding woodlands. Later, in the 1890s, Robinson declared that "the modern garden is often no more interesting than an oilcloth pattern, because instead of beautiful form and colour we see emphasis given to pattern-work and plants robbed of all their grace."[18] He believed that plants should be given the freedom to express their own forms, following their own growth patterns, and not be forced into preconceived compositions, a line of thinking that Migge would take up in *Garden Culture*.

The other major approach that arose as a reaction against the gardenesque and the landscape garden generally was the so-called formal garden, which reflected the desire for another kind of return, this time to the architectural garden before 1750. Perhaps the most vocal and strict advocate of this mode was the English architect Reginald Blomfield. In his 1892 book *The Formal Garden in England*, he described the necessity of emphasizing the relationship between house and garden by following an architectural treatment, with geometric lines and a fixed framework of stone and other durable materials.[19] Predictably, Robinson criticized Blomfield for not giving sufficient consideration to the plants themselves, while Blomfield pointed to Robinson's lack of an overall spatial conception. In historical terms, however, this well-known debate does more to obscure than to clarify, as both were reacting against the gardenesque, albeit in different ways. The more practical, less polemical garden designer Thomas Mawson took a less stringent position, accepting both modes of design in his 1901 book *The Art and Craft of Garden Making*. For Mawson, it was a matter of appropriate relationships, with the most architectural garden areas near the house gradually becoming less defined toward the edges of the garden, eventually blending into the surrounding woodland. Mawson's discussion of the spaces and elements of the country house was systematic and rational, but directed at practical rather than theoretical ends (in contrast to Muthesius). He did not really participate in the formal versus wild garden debate, because he effectively combined both methods within his own designs.

To summarize, during the late nineteenth century in England, the essential move was from passive, aesthetic contemplation of garden scenes toward more active, functional use of outdoor spaces. Although the formal solutions were quite different, they were all directed away from artistic compositions that reformers considered artificial and contrived rather than honest and authentic.[20] The title of Mawson's book was not incidental, emphasizing as it did the connection to the Arts and Crafts movement tenet that an object should clearly reveal how it was made. These new gardening ideas gradually entered Germany in the 1890s, at the same time that garden reformers were ridiculing the German version of the landscape garden as a dead stereotype, with tortuously curved "pretzel paths" and other contrived features.[21] But the resemblance of large-scale architectural gardens to the work of the Frenchman André Le Nôtre (the original designer of the gardens at Versailles) was too close for comfort for the older generation. For them, the formal garden represented aristocratic despotism in contrast to the democratic associations of the "English" garden, which was understood as being something similar to the gardenesque.[22] Thus, the new architectural garden had to be rendered more "German" and traditional in order to gain popular acceptance.

The Garden Reform Movement in Germany: Alfred Lichtwark and Hermann Muthesius

In Germany, Alfred Lichtwark, the Anglophile director of the Hamburg Art Museum (Kunsthalle), first wrote in the early 1890s of the need for a new kind of garden, one based on an imagined archetypal German "farmer's garden." He could thus be thought of as the "father" of German garden reform.[23] Lichtwark's farmer's garden was also a kind of return, this time to the roots of German culture, with obvious nationalistic undertones, yet it was very similar to the cult of the cottage garden in England earlier in the century. Lichtwark believed that the farmer's garden had evolved from the "original garden" (*Urgarten*),[24] meaning the first utilitarian gardens made with simple enclosures. He claimed that, in its more refined form, the plants and the spatial organization of the farmer's garden showed the influence of the medieval cloister garden and, by extension, the classical Mediterranean. While Lichtwark didn't provide illustrations, he explained that the overall division was like a play in three acts: beginning with the gravel terrace near the front door, followed by the geometric cut flower and vegetable gardens, and culminating in a small pavilion with surrounding lawn, the whole connected by an axial path emanating from the front door. All of these areas were thought to be the most appropriate forms for need and use. One of Migge's first garden designs published in 1907 followed Lichtwark's paradigm to the letter, although he soon adopted a more subtle approach. Lichtwark himself later proposed another idealized model, the "heath garden" (*Heidegarten*), so named because it incorporated plants that thrived in the surrounding heath landscape, as a conscious adaptation of the planting principles of William Robinson.[25] Migge mentioned Lichtwark's heath garden in passing, but was somewhat critical of it, as will be seen.

Migge introduced himself to the architect Hermann Muthesius in 1909, and although the two collaborated for a few years, only two joint projects have been substantiated: the Michels Silk-Weaving Factory and the Wegmann House, both included in *Garden Culture*.[26] Undoubtedly, it was Muthesius who conducted the most thorough study of the contemporary English country house and garden, presented in his magisterial three-volume study titled simply *The English*

House, published between 1908 and 1911.[27] Sent by the German kaiser to study English domestic design, Muthesius remained in London from 1896 to 1902, traveling extensively throughout Britain. One section of this work was dedicated to the garden, in connection with new or recently renovated large-scale country houses, many of them by well-known architects associated with the Arts and Crafts movement, such as C. F. A. Voysey and E. L. Lutyens. Muthesius conceived of *The English House* as a documentation of contemporary domestic culture for the German reader. He was particularly interested in the complex domestic program of the country house, from the series of functional rooms attached to the kitchen, to the various reception rooms, to the sleeping quarters. His focus was not on architectural form alone but also on the overall social program that the many individual spaces were intended to serve. Muthesius and the English designers he was considering understood the garden as an extension of the house. Garden plans included by Muthesius show spaces such as bowling greens, tennis courts, rose gardens, covered walks, and service courts, which were repeated in the work of different designers. One garden plan by Thomas Mawson indicated wild garden and woodland areas relatively close to the house, but most of the garden plans selected by Muthesius only delineated the complex of outdoor rooms adjacent to the house, all of these designed to support an active life in the open.

Muthesius organized his discussion in a manner similar to that of Mawson's book, and opened his discussion of the garden with a citation from William Morris also used by Mawson: "Large or small, the garden should look orderly and rich. It should be well fenced from the outer world. It should by no means imitate the wilfulness or the wildness of nature, but should look like a thing never seen except near the house."[28] That the garden should be "organized" was one of Muthesius's main points. He highlighted the kitchen garden as one of the "utilitarian objects which have escaped the attempts at over-refinement of the past century," meaning that the expression of utility was more honest and more appropriate than the artistic effects of the gardenesque. The "regular" (*regelmässig*) was the most functional mode of garden design, and therefore the most suitable for the modern era; it mattered not whether for growing roses or vegetables.[29] Yet on the scale of the large country house, Muthesius was willing to accept a freer treatment of the areas beyond the regular garden, where "the wood is then cultivated as a wood, which means that natural woodland flowers are sown in great profusion in clumps, so that there is always a carpet of blooms to enjoy." According to Robinson and the contemporary English gardener and writer Gertrude Jekyll, such fringe areas were to appear quite simple and untreated, not as a perfected version of pastoral scenery.[30]

In his 1907 book *Country House and Garden,* Muthesius shifted his emphasis from the stately house in the country to the suburban villa on the edge of the city, directed primarily toward the well-off middle classes.[31] Whereas his previous study had been presented as a documentation, this study was clearly an attempt to synthesize a new model for domestic design in Germany. Although he was referring to the suburban villa, he instead used the standard German word for "country house," *Landhaus*, to denote a new kind of house-and-garden paradigm.[32] He criticized the historicist houses then found on the outskirts of most German cities, referring to them as "masquerade villas" because their facades of applied ornaments acted like masks, falsely alluding to a higher social status than that of the actual occupants within.[33] Muthesius probably realized that as a reformer he could reach much greater numbers of people by presenting the typical version of the *Landhaus* at a more modest scale, while still addressing readers with the financial means to realize a higher level of domestic culture. One of the first things that he attacked was the monumental stair of the villa, which he ridiculed by dubbing it a "chicken ladder."[34] The new *Landhaus* was not to stand like a grand isolated object within a naturalistic setting. Rather, the garden should be reached directly from the house interior, he argued, via a terrace, with as few steps as possible. This connection of house and garden was not merely a matter of convenience alone, for as he explained:

> If the house is combined with the garden as a unity, then this can only be the case with a regular [*regelmässig*] garden. The palace [i.e., historicist villa] and grassy lawn are no artistic unity. If the house is architecture, then the garden must also be architecture. And if one takes the word "architecture" in its most general terms, so that it encompasses human creation and design in all its forms, then garden design must belong to architecture by necessity. . . . The same fundamental principles that underlie the house, the same organic relation of the parts to one another, the same unity of the single elements into a harmonious whole . . . must also govern the garden.[35]

Muthesius did not use the German word for "formal," but rather "regular," which for him meant a clear, logical order that was understandable as such.[36] His argument for orderly forms was implicitly a plea for a more spiritually unified existence as well, first connecting the order of the garden to the house as architecture, then to "human creation and design in all its forms," and finally to the order of biology, of life. But for him naturalistic forms were by no means equivalent with organic unity. Whereas the much larger-scale country-house garden could be physically connected to actual meadows and woodlands, he considered the imitation of natural scenery within the smaller-scale suburban garden to be wholly out of place. Muthesius ridiculed features such as a "cement-lined puddle" that was supposed to resemble Lake Lucerne as comic absurdities, for they represented something that they obviously were not, much like the "masquerade villa" with its ornate facade.[37] In the end, Muthesius, as an architect, was less concerned with plantings and the actual treatment of plants and more emphatic about organization and planning.

Migge's "Garden-Architectonic" and His Rejection of the "Ecological" Garden

Naturally, contemporary debates and positions in the field of garden design found their way into Migge's own thinking. In Germany, the battle between the older generation in favor of the landscape garden and the reformers arguing for the architectural garden had been more or less been won by the latter by 1913, as Migge plainly noted in *Garden Culture*. Migge was obviously reading German writers, and while it is not clear to what extent he was actually able to read English treatises, he appears to have been influenced by them; for example, in *Garden Culture* he mentions the work of Gertrude Jekyll.[38] In Anglophile Hamburg, where he was based, it is clear that many among the cultivated were reading Robinson, Jekyll, and others.[39] In any case, the two German writers to have the greatest influence on Migge, Lichtwark and Muthesius, were very conversant with English gardening ideas. Thus, Migge's study may also be understood as part of the wider international interpretation of English gardening discourse. Like Muthesius, Migge considered the large-scale country-house garden design as the ideal, although he aimed much of his discussions at the well-to-do middle classes in their suburban *Landhaus*, which indeed formed the bulk of Migge's own design work. Again emulating Muthesius, Migge also believed that the most appropriate form of garden design was architectural, which he translated as the "garden-architectonic" (*Gartenarchitektonisch*). Muthesius himself had derived his understanding of the "tectonic" from the writings of the mid-nineteenth-century German architect and theorist Gottfried Semper.[40] Following Semper's logic, the building was to be expressed as a clearly perceived assemblage of components rather than as a picturesque composition. Migge logically extended this to the garden, also taking cues from Lichtwark. In one of his more memorable slogan-like statements, Migge declared that "making a good garden is in principle one of the most simple and matter-of-fact things: *one must organize it*." Migge, like Muthesius, believed that architectonic logic was above all orderly and organized, naturally resulting in a good garden.

Unlike Muthesius, however, Migge as a gardener was much more conscious of the role of plants. Migge's garden-architectonic signified a formal synthesis of hard architectural elements and vegetation, as well as the use of plants on their own as architectonic elements to define spaces and to provide structure. He categorized the major vegetal components of the garden under separate headings, moving from the simple selection of trees and shrubs to clipped hedges, then to arbor walkways, allées, and groves, to vine-covered structures, and finally to flowers and flower gardens. The primary distinction that he made between his garden-architectonic and the outmoded landscape garden was that in his approach all plants were to be used to define spaces, not simply to compose naturalistic scenery. On the other hand, Migge repeated previous criticisms of his contemporaries—namely, that plants were often used simply to furnish the architectural garden without giving sufficient consideration to the plants themselves. This kind of garden could even be worse than the landscape garden, for "the intrinsic value of a garden depends upon the particular manner in which we utilize the plants within it."[41] Echoing Lichtwark's interest in the garden as a place to educate the eye and the senses, he continued, "not only plants as the means of fulfilling specific garden needs, but even more so plants for their rhythmic effect among one another: color and form *as such* are decisive here."[42] Migge believed that structure-giving plants should be utilized to reinforce the architectural character of the whole, most clearly in the case of clipped hedges, but that garden plants were also to be valued for their own typical growth patterns. The architectural garden was not composed of strict geometric forms alone, for "we also openly recognize the characteristic beauty of freely growing plants. . . . it is *our eye cultivated in respect to nature*, that appreciates the worth and finest qualities of freely growing plants."[43] The photographic images in *Garden Culture* of Migge's own design for the smaller-scale, suburban Rheinhardt garden illustrate how he was able to skillfully combine both clipped hedges and freely growing shrubbery within an overall geometrically defined plan, both emphasizing spatial divisions and blurring edges through contrasting but harmonious planting treatments.[44]

With a shift to a larger scale, Migge also called for the inclusion of wild flowers, which he claimed had formerly been only "occasionally and timidly used in landscape parks or bosquets."[45] In a manner reminiscent of Robinson, he continued his discussion with an imaginary vision of an "extensive park" with a "softly rising slope with flowers." He then expanded the view outward in his mind: "Farther on, this sea of flowers flows into an area of large-scale beds and agricultural fields . . ."[46] Here, Migge envisioned the inclusion of freely distributed flower plantings not in a "wild" area, but within a landscape clearly shaped by the human hand. While the first impression of Migge's garden-architectonic may be one of rigid control, a more careful reading reveals that he was much more flexible in his use of plants as forms, with the ever-important qualification that they be used to define space, a differentiation from the landscape mode that was perhaps more difficult to express in practice than in theory.

But Migge was also very careful to qualify his position relative to current arguments that gardens should be based on regional landscapes, such as Lichtwark's "heath garden." Migge most definitely was not suggesting that particular kinds of natural landscapes should be reproduced in the garden, for as he asserted, "I fundamentally do not recognize, for example, that a piedmont garden is a type based

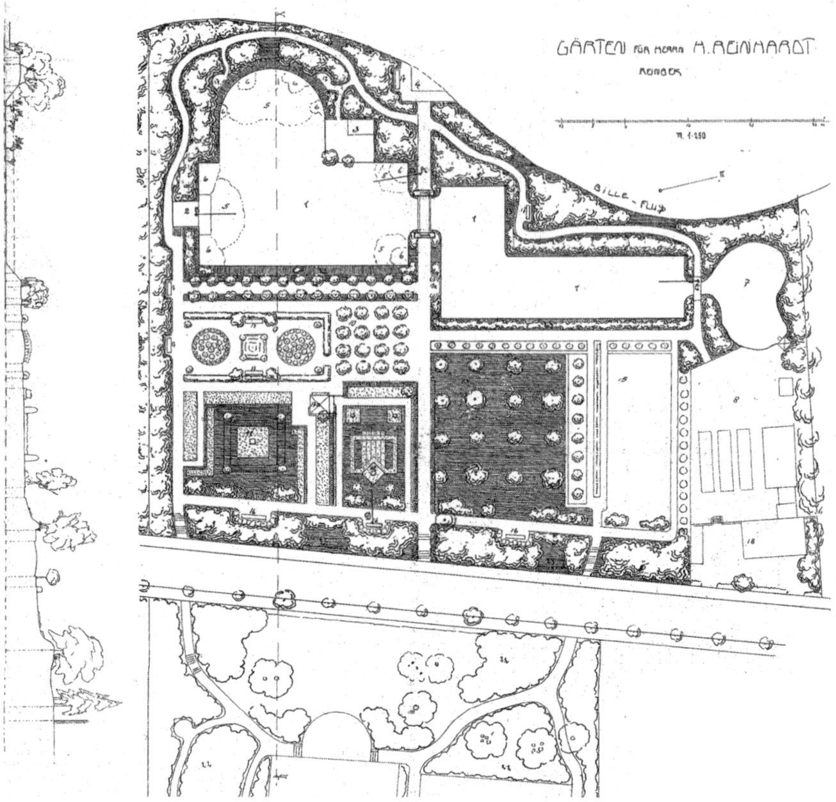

figure 3

Plan of the Reinhardt Garden in Hamburg-Reinbek (1907), by Leberecht Migge. Photographs of the garden were originally included in *Garden Culture*, as Plates 1, 7, 9–10, 15–16, and 22–23.

Reproduced from *Deutsche Kunst und Dekoration* (1913):306.

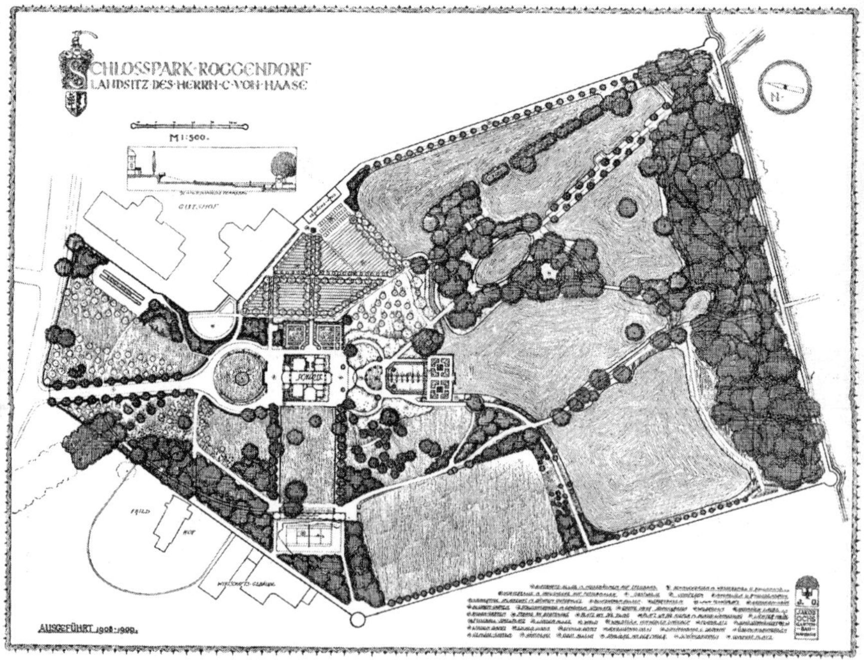

figure 4

Plan of the van Haase Garden in Roggendorf (Mecklenburg). A photograph of the garden was included as Plate 21 of *Garden Culture*. On the plan, to the right of the main house, a series of terraces extended toward the south, culminating in a pair of square terraces. The small dark squares in the centers of these terraces indicate the location of two classical sculptures, one of which is shown in Plate 21.

Reproduced from Leberecht Migge, ed., *Ein modernes Gartenbuch* (Hamburg: Jakob Ochs, ca. 1911), 15.

on local necessity."[47] He tactfully noted that he did not want to "underestimate the potential unique beauty of the 'heath garden' as described by Lichtwark," arguing instead: "But just as with adaptation, stimulating *contrast* can also be emphasized, here with the heath garden motif."[48] For Migge, the significance of a garden in a heath landscape was that it could be achieved at all, and it should be celebrated as such. Migge believed that the garden designer should rely upon the characteristics of plants, their forms, colors, and even scents, and that planting compositions were to be inspired by combinations in the landscape outside the garden, but never copied as a kind of slavish imitation: "Nature is never tasteless. But it is also not always interesting, as we know."[49] Migge's target here was not really Lichtwark, but rather Migge's nemesis, the garden designer Willy Lange, who had published his own much longer book on garden theory in 1907, with a second edition in 1909 and a third in 1912.[50] Lange discussed the possibility of an architectural layout for the garden, but only as one among many options, and not the most desirable one at that. Perhaps the most original (if controversial) section in Lange's book was that on the "natural garden." There, Lange explained that "leitmotivs" in the garden could be based on naturally occurring "plant communities," such as the heath, shore, or mountain landscape.[51]

Like many other thinkers during this period, Lange was trying to base his theories on the objectivity of science, in this case on the relatively new field of ecology, which focused on interdependent biological communities composed of different organisms. Without mentioning Lange directly, Migge bluntly rejected this kind of thinking in the garden: "No, we contemporary people want nothing of garden naturalism. . . . There regularly occurs—a characteristic sign of our current garden rhetoric—a *circulus vitiosus* that moves from nature to science, from reason to art, and back again to nature in a never ending cycle."[52] For Migge, Lange's natural garden was in truth a disguised attempt to reinvent the landscape garden, but with even more pernicious logic. In Migge's version of the new garden, references to nature were a matter of composition, intended to appeal to the human senses. In what may appear to be a paradox, Migge was probably influenced by the popular writings of Ernst Haeckel, the German biologist who first coined the term "ecology" in the 1850s.[53] Haeckel was well known for his drawings of primitive deep-sea organisms and for popular books such as *Nature as Artist*, in which naturally occurring life-forms previously not considered to be "beautiful" were nevertheless shown to be on the level of "art" because they were functional and purposive.[54] In this respect, Migge commented: "We have experienced the new natural sciences: *our eye has grown more subtle*, more calculating, it sees the outward appearance of the plant at the same time as its inner structure."[55] Here Migge turned to a different aspect of biological theory from that which Lange embraced, drawing on human perception rather than a supposed suprahuman ecological unity found in the natural landscape.

Migge's own arguments for the architectural garden followed those of a number of authors writing on the subject over an approximately thirty-year period preceding his own work. But unlike the majority of them (with the possible exception of Muthesius), for Migge the end goal was not just to design a good garden. As he explained, "the transfer of architectural principles onto the garden is to me not merely a formal and not even an aesthetic matter."[56] This approach was a means to an end, but not limited to the *Landhaus* garden as an isolated object. He continued: "No, the architectural design of the garden is necessary for us, because it is so *simple*. Because its elements are easiest to handle and are economical by nature, so that in our era of mass-problems *it alone* results in an effect on a broad scale: *I want the architectural garden for economic, social, and ethical reasons*."[57] With statements like this, Migge moves from the realm of pure garden design, traditionally represented by the large-scale garden and park, to the "mass-problem"—that is, to a systematic approach to garden design that can be applied at the scale of the metropolis and even the nation. He thus led from design reforms conceived to support the new domestic culture, to a wider scale of perceived social needs and objectives, which cannot be understood wholly within the context of contemporary garden design discourse alone.

The German Werkbund and the Garden as Applied Art: Garden Culture for the Masses

Garden reform was related to a range of movements in Germany dedicated to the improvement of the applied arts, which had been expanded to include not only architecture but gardens as well.[58] As a progressive young man who wanted to be at the forefront of change, Migge logically would have wanted to join the Werkbund and adopt its goals. The Werkbund was founded in 1906 with the stated purpose of improving the applied arts in Germany, focusing on standards of mass production and mass consumption. Ostensibly a private organization, it was organized by leaders in many fields including architecture and design, as well as industrialists, economists, sociologists, and politicians. There were two major areas of discussion: the mass-produced object and the domestic environment, the latter in fact being the intended destination for many of these products. But Werkbund members did not limit themselves to their own country, as another primary goal of the movement, which they advocated using the nationalistic rhetoric typical of this period, was to make German products more competitive on the international market. While the economist Karl Marx thought of the mass-produced "commodity" as an abstraction standing for profit, the members of the Werkbund were indeed interested in the characteristics of the physical object, although also with an eye to its marketability. "Economics" and "economical" design for them were not merely about balance sheets and profit making but about the most efficient use of energies and materials, resulting in "good" design, much as biological forms were based on economic principles. In this way, their intentions were directed not only toward an improvement of the physical environment, but also toward "spiritual" elevation as well.[59] These are beliefs that Migge frequently alluded to in *Garden Culture*.

Werkbund members such as Migge and Muthesius thought that new methods of production, including materials and techniques as well as new divisions of labor, had to be reflected in designed objects and in the designed environment. As part of the general move from the fine arts to the applied arts, it was thought that artistic expression should directly reflect all aspects of making. Among others writing on this topic was the politician Friedrich Naumann, who expounded, "what has changed are *the clients*, what has changed are *the artists, the craftsmen*, as well as the *production methods*. . . . either a new art will be awakened by these completely changed conditions, or there will be no art for us at all."[60] Naumann's statement was, on the one hand, a rejection of an exclusive return to old-fashioned handcraft methods, and, on the other, a call for everyone in the applied arts to become actively involved in the processes of industrial production. This set of relationships is probably clearer in the case of the industrially produced object than with the domestic environment. But as Migge realized, the conditions of production also affected how the garden was produced, both through plants as "material" and through the division of labor between the intellectual and the physical. During this period, "landscape architecture" was still not firmly established as a profession, for divisions between the intellectual, artistic, and physical labor of producing a garden were still being defined.[61]

In this respect, Migge's book is again unusual for a garden treatise in that he has a section titled "People and the Garden," in which he considers everyone involved in garden production: "When the agents here in question—the gardener, the architect, and the layperson—come together with a proud and eager spirit, at whatever cost, to achieve gardens worthy of our day: *then everything is possible!*"[62] Migge believed it was necessary to educate not only the designer but also the consumer. The garden, like the house, was a kind of *Gesamtkunstwerk*, or total work of art, where products such as garden furniture came together with hard elements produced on site in materials such as cast-in-place concrete (then a new technique). These elements were combined through the garden-architectonic approach with plants raised elsewhere, often in large-scale horticultural facilities. Migge also discussed plant producers as another important human component, for he believed that they must be enlightened in the same way as factory owners in order to elevate their production standards and aesthetic judgment.

Changing conditions of production also involved new materials and new ways of utilizing materials long in use. The "honest" use of materials, as opposed to manipulating one material to represent another, became another theme within the Werkbund. Here the garden reformers' rejection of elaborate planting compositions such as carpet bedding clearly intersects with Werkbund thinking. As Migge explained his own terms, art began with the "*selection of materials*."[63] He acknowledged the difficulties of applying standards for material selection from other fields: "The translation of an idea into being, from credible concepts to tangible material, is a fundamentally different process in applied horticulture than in the otherwise closely related areas of architecture and interior design. . . . Plants as a material are too mutable and too unstable."[64] Thus, Migge sensibly recognized that there were other factors coming into play in the production and maintenance of gardens. He was also careful to point out that botanical knowledge alone was not sufficient for "rhythmic garden design," any more than "the knowledge of concrete production is for the attractiveness of a house made from this material."[65] At the same time, he celebrated the transcendent quality of the garden: "*It is living*. Everything wakes and grows in it and is not dead like cold stone and hard timber."[66] Rather than see the mutability of garden materials as a problem, Migge associated the living quality of plants with the Werkbund desire to reconnect design to the spiritual order of life.

One of the main theoretical terms proposed within the Werkbund was the "type," most remembered as the subject of a debate in 1914 between Muthesius, as its most vocal advocate, and Henri van de Velde, representing the opposition.[67] The Belgian architect Van de Velde believed that the type, as a concept based on repetitive form, restricted the creative expression of the individual artist. But the "type" as a concept could be traced back to Platonic and Aristotelian analyses of "form" and "idea."[68] More recently, the type had been introduced into architectural discussions in the mid-nineteenth century by the German architect Gottfried Semper, who took the term from biological studies showing that organisms evolved from elemental, typical forms.[69] In terms of mass production, Muthesius and others thought of the type as a basic idea of form that could be repeated with variations without losing an identifiable appearance. The series of teakettles designed by Peter Behrens for AEG (the German equivalent of General Electric) shows how a basic form could be repeated with subtle variations.[70] Ultimately, the Werkbund conception of the type was a means of limitation and control, since thinkers such as Muthesius believed that the individualism of art movements such as impressionism and *Jugendstil* (Art Nouveau) had corrupted the purity of architectonic expression. The type would assure adherence to forms that not only were tasteful but also assumed a kind of universal core value. More than this, the type represented an inherent unity that in its essence represented a higher plane or order. Defining the type as a working concept for the design of mass-produced objects proved to be much more difficult than postulating an abstract theory.

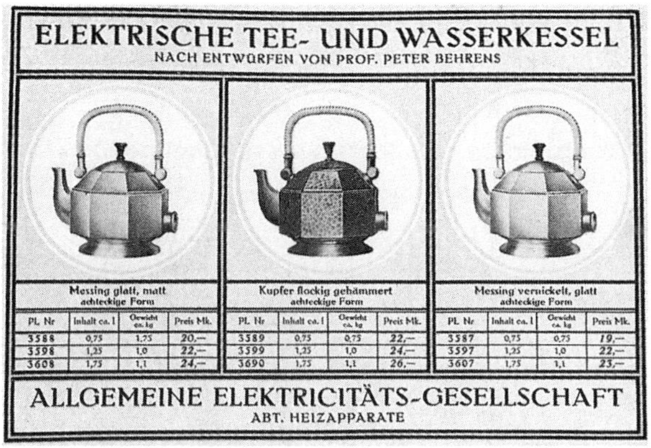

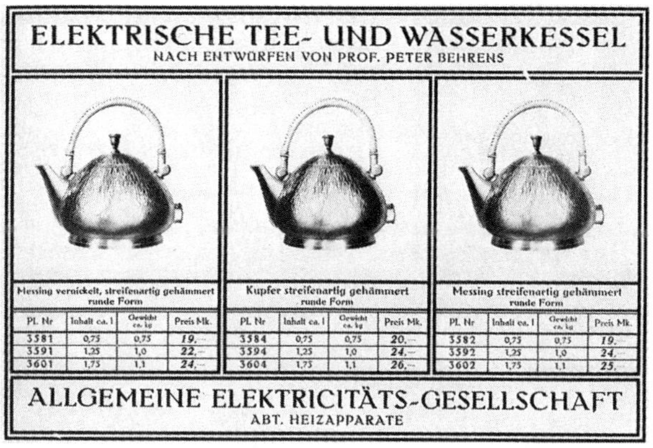

figure 5

Advertisements for a series of tea kettles designed by Peter Behrens, illustrating the concept of the repetitive form type, in this case as a mass-produced object.

Reproduced from *Tilmann Buddensieg, Industriekultur: Peter Behrens und die AEG, 1907–1914* (Berlin: Gebr. Mann Verlag, 1979), 427–28.

Migge was highly enthusiastic about the idea of the type, and as both a garden designer and a writer he struggled to incorporate this principle within the field. He thought that the type was necessary because the intellectual and creative capacity of the German people could not increase proportionally with the population but would remain in the same measure. Therefore, the genius of the designer could not be reserved for the few but must be available to the many: "*Such a thing that is conceived for the many simultaneously, we call a type.* We mass people need types."[71] As historical backing for these statements, Migge then seems to have returned to Lichtwark's analysis of the farmer's garden, proposing that "the utilitarian garden can be categorized as the first typical garden. All the garden art of earlier periods indirectly arose from this experience. Thus, the later ornamental gardens are nothing other than the spiritual and material elevation of the vegetable gardens, herb gardens, and orchards that preceded them."[72] With this example, he wanted to show that later garden forms could be traced back to an original one and that their evolution occurred through a progressive sequence, his logic following the contemporary tendencies equating biological and historical development. He and others believed that the landscape garden was an unfortunate deviation from this sequence, as it represented a diversion from the true character of gardens based on the act of gardening. A return to the true evolutionary pattern was necessary for the future progress of garden design, or so he believed.

For Migge, this historical evolution was not a completed process: "Not only have we embraced the arbors, allées, and groves, the pavilions and pergolas of earlier garden periods, and adapted them to our needs, in addition many new forms are also appearing."[73] The evolution of garden types also involved recombinations of simpler components, such as those listed above, into more complex wholes that would further evolve and overlap. In Migge's design method, the spaces in the garden, as types, could be reassembled under the given conditions of a particular project, almost like a kit of parts. This understanding was also in keeping with the tectonic approach of articulating all of the physical components within an overall assemblage. Migge applied this principle even at the smaller scale of shapes in flower beds, for the flower garden plans included in *Garden Culture* were also conceived as designed objects composed of smaller elements that could be recombined.

On another level, the domestic program that Muthesius had identified in the country house could be assigned within Migge's garden culture program to a series of "typical" spaces, which he listed as the garden terrace, the communal lawn, the flower border, various play areas (such as tennis courts and football fields), and even sunbathing and swimming facilities. Migge also organized the discussion of whole gardens as types under headings such as "the concert garden," "soldiers' and railway gardens," and "the factory garden." While many earlier garden treatises had categorized gardens by use, Migge implied that new, evolving uses demanded particular forms, which could only be effectively realized through syntheses of garden types. At the same time, he recognized that this method of design could be criticized for curtailing creativity, a charge he defended against: "Is this a lack of imagination? No, it is a restraining tendency, a self-reflection, and a limitation to the essential. It is called *organizing*."[74] Throughout *Garden Culture*, Migge returned to the theme of organization, which clearly meant more for him than simply establishing practical relationships. "Organization" here is rooted in the concept of the "organic," in the older sense of the term, meaning that the component parts should naturally relate to the whole, as in the word "organism." The type and the typical were thus not just patterns or even categories of use but entities that were part of a historical evolution of forms that arose as rational, orderly responses to function.

Migge's adaptation of the type is less convincing when he tries to assert that his garden culture is like a discrete product that could be exported, following the Werkbund aim of improving Germany's position in the international market. This was of course an economically sensible move on the part of the Werkbund, but the level of bombastic nationalistic rhetoric employed in this respect may now seem somewhat appalling, as demonstrated by the tone of the conclusion to *Garden Culture*. While Migge claimed that the production of goods and materials for the garden, as well as labor within the garden, had economic significance, he did not clearly explain how German gardening knowledge was to be internationally exported. This did not seem to be a problem for him, for he proclaimed that, "Here there will be conquerors and conquered. But Germany doubtlessly holds the best position."[75] He wanted to promote garden culture, just as any other product for export: "New gardens: 'made in Germany'!"[76] (Note that "made in Germany" appears in English in the original, for emphasis.) However valid this argument may or may not have been, his comments reinforced popular nationalistic sentiments common in Germany on the eve of World War I.

The discussion of spirit and the respiritualization of artistic production was largely drawn from the popular reception of the work of Friedrich Nietzsche.[77] His philosophy was virtually institutionalized through the Werkbund and other organizations, where Nietzschean concepts maintained an important background role.[78] Nietzsche had rejected the old ideal of disinterested aesthetic contemplation formulated by Kant, according to which the aesthetic object was thought to have a relative degree of autonomy (as with a painting in a museum, or a vase in a glass case).[79] Following Kantian thinking, applied art was not art as such because it had a functional use value. As a direct rejection of this position, Nietzsche embraced art as part of life. Art was not to be separated from life but completely integrated within it, both in the biological sense and in the sense of day-to-day living. Art was in fact to draw its meaning from life, or it was not to be considered art. At the same time, Nietzsche called the appearance of forms into question, for forms themselves did not represent truth as such. In the Werkbund context, this was interpreted to mean that physical objects resulted from underlying social and economic as well as biological life forces. In order for the object to be elevated to a higher spiritual level, the artist or designer had to be consciously aware of these forces and to express them.

Another critical tendency that affected the thinking of cultural reformers such as those in the Werkbund was the "critique of civilization" (*Zivilisationskritik*), aimed at post–Industrial Revolution society. Such criticisms were often nationalistic, arguing for the renewal of a specifically German way of living and being, in opposition to "liberal" thinking, blamed largely on the French Enlightenment. One influential thinker in this vein was Ferdinand Tönnies, who formulated the opposing terms "community" (*Gemeinschaft*) and "society" (*Gesellschaft*), the former referring to the traditional, holistic, and thus healthier form of human organization, and the latter to the alienating, depersonalizing, and mechanistic effects of uncontrolled capitalism and laissez-faire liberalism.[80] Among laypeople—that is, among elitist reformers—this was combined with the Nietzschean search for the truths underlying form, together with a program of social and spiritual renewal through art and design. Again, these were also linked to a defense of Germanic values, and by extension of German competitiveness in the world, both philosophically and economically.

The title of *Garden Culture* is not insignificant, for Migge was clearly trying to link his gardening system to the wider critique of civilization, which aimed for a reestablishment of ethically grounded culture. Like other Werkbund writers, he recognized the importance of industrialization and capitalism to modernity, but he also believed that they should not be allowed to dominate; order and spirituality should still be actively cultivated within mass culture. Cultural reform meant probing beneath surface appearances in order to understand underlying forces. This belief was clearly revealed in one of Migge's opening statements, that "first we must sweep aside the great rubbish and detritus that covers over the field of gardening,"[81] referring of course to the landscape garden and its proponents. He continued: "This complicated garden-web that now enmeshes us . . . only signifies a veil over the chaos that spreads beneath."[82] Here Migge layered Nietzschean philosophy over the more typical garden-reform rhetoric. In comparison to previous arguments in England in favor of the architectural garden, Migge's version was much more theoretical, embedded as it was in the popular reception of German philosophy.[83]

Some of Migge's phrases in *Garden Culture* were even more obviously Nietzschean in derivation, such as the "will to garden culture" (*der Wille zur Gartenkultur*) or "garden will" (*Gartenwillen*).[84] These were similar to the phrase "will to form," employed by other critics of the day. In this latter sense, Migge's "garden will" may have also been derived from Alois Riegl's term *Kunstwollen*, or "artistic will," interpreted by Muthesius as the essential, driving force underlying artistic creation, in opposition to mere historicist copying of external stylistic appearance.[85] Cultural reform was opposed to dry academicism, oriented instead toward action, toward doing and creating, again themes found in Migge's writing. Another aspect of Nietzsche's thinking was the rejection of the purely rational, which reformers interpreted as a call for the cultivation of the spiritual and even the mythical. "Garden will" for Migge signified a collective will, not only as a mass-psychological movement, but also as a spiritually oriented recovery.[86]

Migge's lengthy, sarcastic discussion of the various titles that garden designers were then using, such as "garden artist" (*Gartenkünstler*), is related to the rejection of the garden as art in the sense of the passive, aesthetic object. He probably coined the term "architect for horticulture" (*Architekt für Gartenbau*) for himself, as an indication of his claim to a more objective, active position. Migge repeatedly used the German term *Gartenbau*, which can be translated as either "horticulture" or "gardening," and he probably meant it in both senses, the latter implying a comprehensive "garden culture." At the same time, architects such as Muthesius were using the term *Baukunst*, which literally translates as "building art," rather than the Latinate term *Architektur*. *Bau* was preferred as an ancient Germanic word root, used in both cases in part to emphasize a connection to the origins of Germanic culture.

Another term from Werkbund discourse that Migge used repeatedly was "rhythm," also employed frequently by Muthesius. As with the turn to biology as a means of scientifically grounding design, here "rhythm" was taken from physiological psychology and was thus thought to be an objective principle. In *Outlines of Psychology*, Wilhelm Wundt, one of the most important writers in that field, explained the importance of "rhythm" as a fundamental principle of human expression based on intervals of time.[87] Wundt considered rhythm to be the primary principle regulating movement in music and dance, and by extension the regulation of implied movement within visual compositions as well. Rhythm could thus be simultaneously experienced through different perceptual modes, not only through sight and hearing but even through the entire body, as in the case of dance or the promenade through the garden or park. In this way, different forms of artistic expression could be integrated through rhythm, much as a *Gesamtkunstwerk*. In *Garden Culture*, however, Migge primarily appeals to rhythm as a visual phenomenon, directly in relation to form and color in planting compositions, and more abstractly when referring to a garden or park as being generally "rhythmic." Essentially, rhythm for Migge was the principle relating components to one another within an overall composition. He argued that by focusing on their innate sense of rhythm, those who created and maintained gardens would assure that good designs could be achieved, in accord with human perceptual need.

Migge reveals his understanding of rhythm as a connective, harmonizing principle when he states, "I cannot imagine a perfect park without a definite idea, without rhythm, that would in a certain sense bring this multitude of things into one line, leading toward a monumentality."[88] He also attached specific meanings to rhythm, in passages such as: "The regularity and control of the geometric line promises a stronger, more rhythmic effect in opposition to the arbitrariness of the free."[89] While rhythm was not necessarily understood to be "regular" or "controlled," Migge asserted that these were more clearly perceived rhythms and thus superior to the "arbitrary," referring to the freer compositions of the landscape garden. Obviously, Migge's claim was not objective, despite his desire to couch his arguments in allegedly scientific terms. The most notable exception in Migge's use of rhythm in this book is found toward the end: "*This will be brought about through the rhythm of work, which arises out of the varying, constant utilization of all small spiritual impulses and real actions that together constitute stimulating garden living:* this is the true rhythmic basis of the garden culture of the twentieth century!"[90] Migge probably took the phrase "rhythm of work" (*Arbeits-Rhythmus*) from the economist Karl Bücher, who observed that through the rhythms of accompanying song and music, work could be relieved, and the motions of the less skilled more effectively coordinated.[91] While Bücher as an economist was concerned with increased production and better working conditions, Migge suggests that the "work" of creating and maintaining a garden through gardening as an act could also be guided by our innate sense of rhythm, our "small spiritual impulses." The most obvious means of discussing rhythm and movement would seem to be through physical movement in the garden, but Migge never applied the term directly in this sense. The idea that the rhythmic forms of the garden would channel human movement was only implied.

Indeed, although Migge and others within the Werkbund claimed that their designs were based on new, more objective principles, they were often somewhat vague as to what they actually meant. Migge's use of the "type" is not so very different from that of the "categories" that had long been employed in garden treatises, such as that of C. C. L. Hirschfeld.[92] The type was supposed to be based on universal forms, yet even Muthesius was frustrated in his attempts to formulate a clear definition that could be easily applied to mass production. Migge only avoided this problem by alleging that spatial types were architectural and geometric, thus relying on specific forms that could be readily identified and repeated. Attempts to ground aesthetic perception on more objective principles taken from scientific observation, such as equating rhythm with architectonic form, were similarly fraught with conceptual difficulties. Garden reformers such as Migge decried the landscape garden for being merely an object of passive aesthetic contemplation, rather than places for healthy, active use. But Migge himself designed areas in the garden for aesthetic contemplation, even if based on "modern" principles of perception. On the whole, Migge and other reformers during the period may have succeeded in creating new forms of expression, but their attempts to prove the superiority of their position over the preceding generation through an appeal to scientific principles were problematic at best.

The Metropolis as Mother of All Gardens: Urban Green Space Planning and the Type

In the second chapter of *Garden Culture*, "Metropolis and Garden," Migge set the scene for the location of all of the garden creations that he wanted to discuss: the big city or metropolis, as the primary locus of activity in capitalist, industrialized society.[93] In this context, Migge intended the word "garden" to be understood in a generic, all-encompassing sense, similar to the phrase "green space" today. Gardens were not just an ameliorating factor in urban life, for the city itself was the generating force, as he exclaimed: "*The metropolis a mother of gardens!*"[94] While many thinkers taking up the critique of civilization (*Zivilisationskritik*) wholly condemned the industrial city as the symbol of all that was wrong with contemporary society, for Migge and other reformers the city was to be appreciated for its inherent qualities. The big city was to be improved to support life and not thwart it. This position was similar to that of the Werkbund in regard to industrial mass production; it was a matter of applying good standards of design and planning to elevate mass culture. Migge's garden culture can only be understood against the background of the metropolitan condition. Other writers were celebrating the metropolis in a similar way, such as the architect August Endell who wrote of the special aesthetic qualities of the big city, arguing for its positive potential for human perception.[95] The critic Karl Scheffler relied upon biological analogies in his analyses, claiming that the city was "something living," whose pattern of growth could be "understood as the natural history of the city."[96] Migge himself based his arguments on public health and economics, for he believed that "gardens" or green spaces of all kinds were needed for physical recreation and overall hygienic conditions.

Migge opened his discussion of "public gardens" by lamenting the lack of "garden statistics" and of government policy on the provision of gardens and green space.[97] In response, he furnished his own statistical tables at the back of *Garden Culture*, indicating the ratio of public green-space areas per inhabitant, and categorizing the areas according to type and city size. He also included a table of major English cities, to show just how deficient the German ratios were in comparison, as a means of goading German leaders to action. The inclusion of gardens in the metropolis was also a matter of spatial organization at an urban scale: "It is not only good and practical that all of these garden forms are available for the metropolis, it is also necessary that they are intelligently distributed throughout the urban fabric following a carefully determined, all encompassing *general plan*."[98] In 1913, urban design (*Städtebau*) was still relatively new as a professional discipline, and boundaries with other fields remained to be defined. As discussed above, Migge was certainly not the only garden designer to write extensively about urban planning and parks during this period. Thomas Mawson in England published a treatise on town planning in 1910, and J. C. N. Forestier in Paris had penned a brochure on large cities and park systems in 1906. The move from the urban park to the park system, and then to urban planning, was a logical one for garden designers, who were then eager to put forward their own ideas. Perhaps what distinguishes *Garden Culture* from other works is that Migge wholly integrated the private garden and the public garden (or park) within an overall system.

As mentioned earlier, the single most important source of Migge's ideas on urban planning was the economist and planner Werner Hegemann. The two men were similar in age, and probably met through their senior mentor Muthesius. Before World War I, Hegemann traveled extensively through the United States, making connections among American planners and landscape architects, and studying urban conditions there firsthand. The new Plan of Chicago proposed by Daniel Burnham and his colleagues aroused great interest among Europeans, including Hegemann and garden designers such as Mawson and Forestier. The 1911 publication of *A Park Book*, by Hegemann and Migge,[99] had the significant subtitle *American Park Facilities*. Indeed, the primary focus of *A Park Book* was the Chicago Plan and the older Chicago park systems, along with the Boston metropolitan park system and the older inner parks by Frederick Law Olmsted Sr.[100] European planners and designers were captivated by the enormous scale of park planning in the United States. Hegemann published photographs and plans in *A Park Book*, including a plan of Greater Boston, to which he added an equivalent-scale plan of Berlin's Tiergarten (the "central park" of that city) as a comparative device intended to show how much more green space the American city possessed. In fact, Migge used many of these illustrations in his own book, including the Boston plan.[101] Hegemann emphasized the different character of the parks in Boston depending on location, beginning with the urban center then moving outward to the regional or "metropolitan" parks, which were essentially large, improved tracts of countryside.

That Hegemann's analysis of the Boston park system exerted a strong influence on his colleague is revealed by Migge's observation that once a general plan had been established, "then only in the correct location will an urban green space obtain its optimal effect, regardless of what kind it is, and only the most direct connections to other similar spaces guarantee the most intensive use."[102] Migge applied the idea of the type to the distribution of these green spaces; here, the type was appropriate not only according to use but also location. Broadly, his system of spatial types began with the urban square, moving outward to the inner park, then to the outer park, and finally to what he termed "open spaces" (*Freiflächen*) on the edge of the city, based on the Boston metropolitan parks. In terms of design, he described the urban squares as being the most architectonic, and the outer open spaces as being the least defined and designed. Although he did not express it as such, this system was not dissimilar to the zones of the country-house garden and park, beginning with the most architectonic area near the house and moving outward toward the woodland garden on the edge of the estate. The metropolis acted as a spatial matrix within which these green-space types were related to each other as components within an overall composition, almost as a very large-scale version of the country-house garden composed of smaller subspaces.

Within his system, in which every green space was considered as a kind of "garden," Migge divided these among three inclusive categories: "those reserved for the use of *individuals* (private), followed by others that serve the *community*, and finally a series of *garden-like* forms found in the metropolis, such as the cemetery or zoological garden."[103] He began his discussion of the private garden not with those of the wealthy but with those of the less well-off. At the smallest scale, this began with the allotment garden, ranging from those provided to aid the poorest of the poor, to allotment-garden colonies leased by workers, to the so-called bourgeois garden (*Bürgergarten*) of the solid middle classes.[104] Allotment gardens were already well established in Germany and often referred to as "Schreber" gardens, named for a mid-nineteenth-century Leipzig physician who had campaigned for communal spaces where working-class children could exercise outdoors.[105] Thus, Migge was capitalizing on an already popular movement, rationalizing it into a system within his overall garden culture program. From the allotment garden, he moved on systematically to the more ornamental front garden, to the larger suburban back garden, and then to his final destination: "A few more kilometers farther out we come to the region of the *Landhaus gardens*. . . . Here is the El Dorado of the garden enthusiast, here he can indulge his passions practically unhindered."[106]

Within his overall scheme, Migge discussed the private gardens and parks of the wealthiest classes in terms of general facilities provided for different kinds of sports and physical activities, as well as for flower cultivation and passive contemplation. In his mind, there was no contradiction in a system whereby the gardens of the poorest and those of the wealthiest could be objectively treated as design problems within a continuum, for "the people should not envy the park owner, but must instead *observe* him."[107] How exactly this was to be achieved in the case of private parks was not clarified, but his statement reveals the kind of top-down paternalistic mentality held by Migge and other reformers during this period.

The first of the public gardens considered was the urban square, which Migge claimed was symbolic rather than functional, its ornamental plantings being of little functional use. Here, Migge echoed the observations of Camillo Sitte, an Austrian designer who wrote one of the definitive texts on urban design in 1889.[108] Sitte believed that such urban green spaces should act as "sanitary" rather than decorative green, by providing healthy breathing and resting spaces. Migge personified the ornamental urban square as being a "lazy" member of the community, who should be made to serve a more active role by providing refuge within the busy city.[109] In his general discussions of public parks, Migge condemned the old-fashioned landscape park along much the same lines as in the section on private garden design. Here, however, the provision of facilities for active use was for the public at large, not the private owner. A concern with health and the human body is clearly evident. Migge first attacked the usual custom of including a beer garden by sarcastically labeling the old-fashioned urban park as the "German beer park," condemning the consumption of alcohol as being antithetical to the purposes of parks—not uncommon rhetoric for the period. Just as with the private garden, he also attacked the aestheticized emphasis of beautiful specimen plantings in the public realm: "Our masses don't want shrub and tree museums in the park that belongs to them; they rightly require facilities that are for active use, not just for looking at. . . . we don't need *Sunday gardens*!"[110]

This statement is intended to be anti-elitist on one level, as garden reformers such as Migge considered the ornamental public park to be an expression of the upper-class preference for wealth and display, here imposed on the public at the taxpayer's expense. He argued that more simply designed functional public spaces were less expensive to realize and maintain, pointing to his own recent public park design in Oldenburg as evidence. Whatever the paternalistic tone he and other reformers may have used, he by no means thought of himself as an elitist: "The ideal people's park must return to every social class what the big city has taken away from them. *The socialization of the public green is one of the most important tasks of our time; it is no longer to be delayed.*"[111] While this call to action may at first appear exaggerated, to judge from the tables that Migge provided showing the ratio of urban green-space areas per inhabitant in major German cities, he probably had good cause to urge for political agitation.

Migge referred to the public parks on the outer fringe areas of the metropolis as "open spaces," where the landscape was least manipulated and most "extensively" maintained as meadow and woodland. This zone offered the most freedom for movement, permitting activities such as skiing and air sports that required large-scale spaces. Migge argued that these open spaces were not a burden on the public purse, but that they acted as a "safety valve"—another hint of elitist concerns over mass unrest.[112] Although Migge claimed that the concept of open space was only beginning to be known in Germany at this time, the Duchess Dohna-Poninska had already written in the 1870s of the need for open spaces on the outskirts of the city,[113] and the urban planner Richard Baumeister had soon after repeated this theme in his own treatise.[114] In fact, during that same decade, the Corporation of the City of London purchased large areas on the outskirts of the metropolis, such as Epping Forest, for the purpose of providing just the kind of open space that Migge later described.[115] Reform parks had already been designed in Germany before Migge published *Garden Culture*, with Schillerpark in Berlin, designed by Friedrich Bauer in 1907, being the first recognized as such.[116] But Migge's study differs from the publications of many other landscape architects of the time, such as Mawson and Forestier, because he considers private and public green spaces together rather than as separate categories, following his belief in the reciprocity of formal types to meet all needs.

This kind of design logic is perhaps even more obvious in the following section on "garden-like" forms, or garden types for particular institutions, all of which are essentially variations on the large-scale country-house garden and park. Migge again describes how his system can accommodate different functions as needed, while still bearing a kind of "typical" or familial resemblance. Unlike public park spaces, these garden-like forms were not discussed as having any particular location within the metropolis, here again having more in common with private house grounds. The cemetery is perhaps the most illustrative case among these, for the inherently repetitive unit of the grave site allowed Migge to discuss the principle of the type most convincingly.

In the final section of the chapter on "Metropolis and Garden," Migge addressed the "garden city," then a relatively new idea coming from England. The garden-city concept was developed by the English reformer Ebenezer Howard in the late 1890s, and not long afterward the first garden-city association outside of England was founded in Berlin in 1902.[117] Migge participated in an official garden-city tour of England in the summer of 1910, visiting Bourneville, Hampstead Garden Suburb, Letchworth, and Port Sunlight, as well as the cities of York, Liverpool, and London, among other places.[118] In *Garden Culture*, Migge, as a skilled gardener and horticulturalist, approached the garden city primarily from a technical standpoint. For him, the main purpose of the garden city was to provide real gardens, not just as decorative green open spaces but as places of intense activity. Thus, he recommended that above all, residents should be trained in effective gardening techniques and encouraged to make gardening a central facet of garden-city life. He did not ignore the design aspect, however, as he described in detail how a full outdoor domestic program could be compressed to fit into the smaller individual garden spaces of the garden city, and in respect to the community as a whole how plantings could be used to create harmonious street scenes. Within five years, at the end of World War I, Migge turned his energies toward the small, self-sufficient settlement based on the garden city, but with even more emphasis on intensive gardening technology, as a means of providing relative economic freedom for inhabitants by helping them to raise their own foodstuffs.[119]

Additional Park Projects: Leipzig, Rüstringen, and the Forgotten Hamburg Sketches

Garden Culture was already in the process of being published when two of Migge's most important public park designs were completed: the Mariannenpark in a suburb of Leipzig, and the main city park for the new town of Rüstringen.[120] He would have considered both of these to be "outer parks" following the logic of his system,

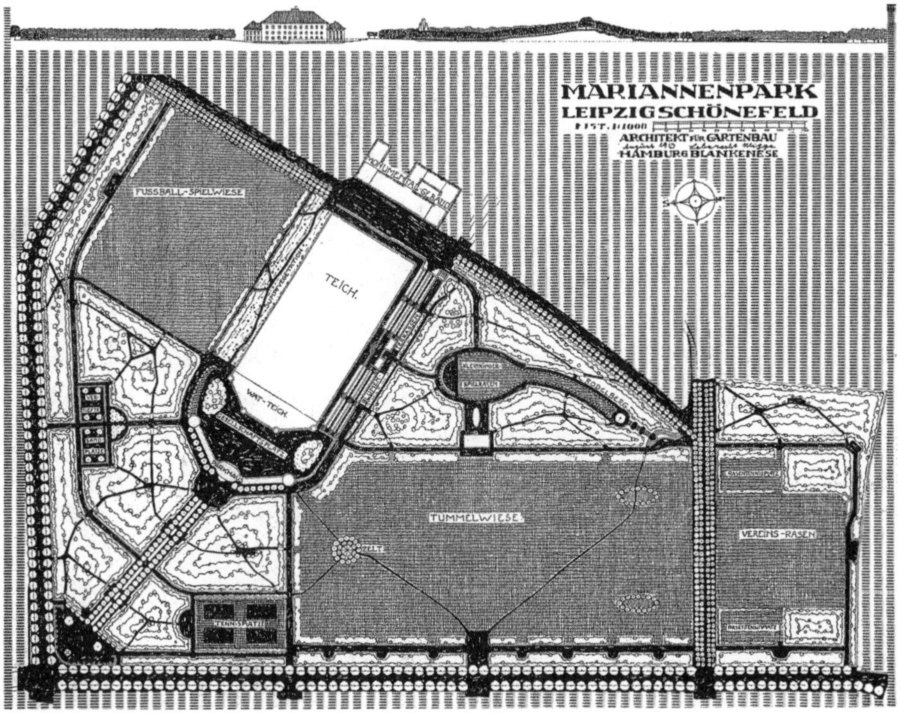

figure 6

Plan of Mariannenpark in Leipzig-Schönefeld (1913), by Leberecht Migge, showing biological figures such as the sledding hill (*Rodelberg*), a plan diagram of the male reproductive organ, and the communal square (*Gesellschaftsplatz*), a plan diagram of the female reproductive organ.

Reproduced from Gustav Amman, "Neuerungen bei öffentlichen Parkanlagen," *Die Gartenkunst* 12 (1914):191.

figure 7
Aerial view of Mariannenpark.
Courtesy of the heirs of Leberecht Migge.

and no doubt would have included them in his book. The Leipzig park was the largest he had designed to date, the individual spaces within it enlarged to the point of becoming large fields rather than outdoor rooms, suggesting a kind of protomodernist open space minimally defined by bordering plantings. At this point in his career, Migge succeeded in avoiding any references to historical versions of the architectural garden; the spaces and forms here were truly abstract. A close inspection of the plan reveals somewhat naive graphic representations of male and female reproductive organs, undoubtedly inspired by popular biology and life reform arguments for increased freedom of the human body. These were merely fanciful plan figures, however, which were not perceptible three-dimensionally. Period photographs and planting plans reveal that Migge applied his planting concepts to animate the austere spaces here, not through elaborate compositions but through the forms and colors of the plants themselves, which were used to express the "rhythmic intentions" of the design.

The second park was located on the outskirts of the city of Rüstringen, which had been recently created by amalgamating three existing towns, primarily with the intent of serving the rapidly expanding German navy. (It is now the city of Wilhelmshaven.) In this former low-lying agricultural landscape, Migge used existing features such as hedgerows to define spaces, adding a long canal to act as a spine uniting the whole composition, and providing necessary drainage. Although he did not claim this himself, he may well have been influenced by the writings of

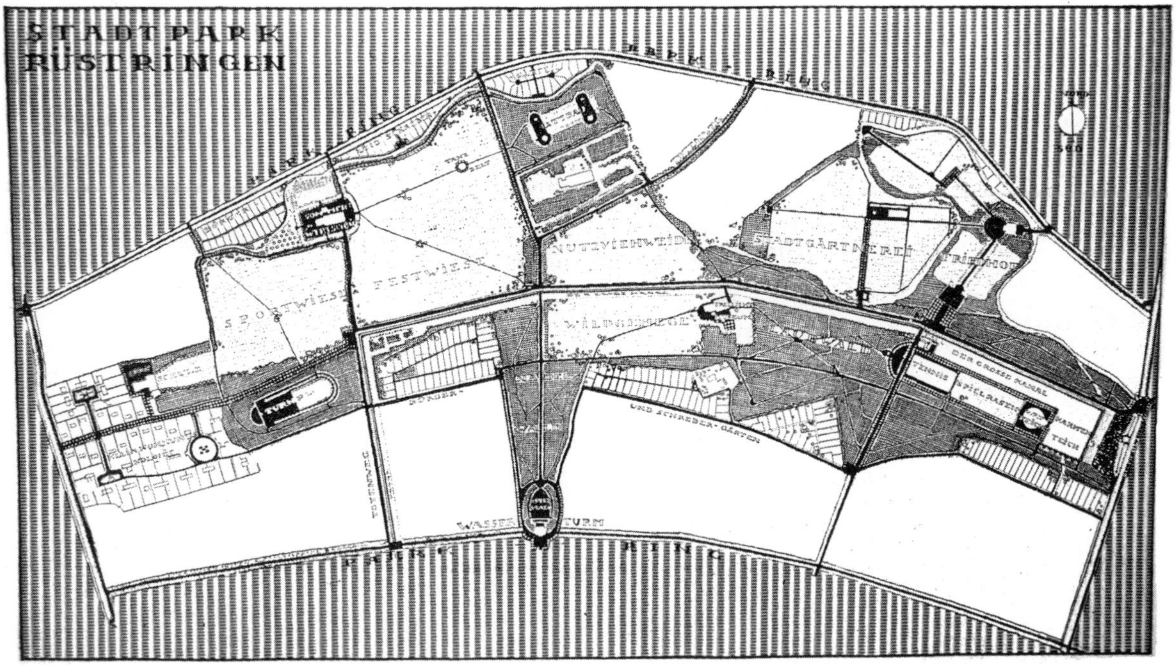

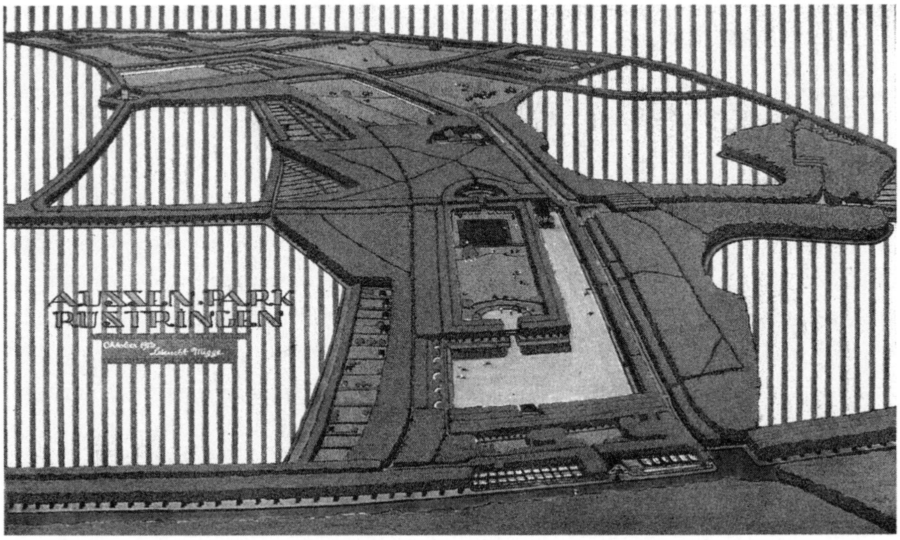

figure 8 (top)

Plan of Rüstringen City Park (now Wilhelmshaven, 1913), by Leberecht Migge, showing a spine-like canal form and the minimal changes in outline distinguishing open space from surrounding farmland.

Reproduced from Gustav Amman, "Neuerungen bei öffentlichen Parkanlagen," *Die Gartenkunst* 12 (1914):183.

figure 9 (left)

Aerial view of Rüstringen City Park.

Reproduced from Gustav Amman, "Neuerungen bei öffentlichen Parkanlagen," *Die Gartenkunst* 12 (1914):preceding p. 181.

figure 10
Unpublished watercolor sketches of a previously unknown proposal for the Hamburg City Park (1909), by Leberecht Migge.
Reproduced courtesy of the author (top) and the heirs of Leberecht Migge (bottom).

one of his role models, Prince Hermann von Pückler-Muskau, who in the early nineteenth century had written of the desirability of making only the most minimal changes to the contours of the land as needed to emphasize inherent characteristics.[121] As an "outer park," the Rüstringen design was only slightly more refined than the surrounding countryside, which Migge would have seen as potential open space for the future city as it grew. In both cases, these parks were designed to be as functional as possible, and indeed Migge won the competition for the Rüstringen park because his was the most economical scheme to execute.

A third design (ca. 1909), for the new Hamburg City Park, was never executed or published, and only two watercolor sketches of Migge's proposal remain to provide evidence of another direction in his thinking.[122] He never received this commission because the newly appointed city architect, Fritz Schumacher, soon assumed control of the design of the new park, now one of the most important examples of the period.[123] As Migge wrote in *Garden Culture*, he believed that plants could be allowed to grow freely, expressing their own will and form, provided they were consciously employed to effectively define spaces—clearly the approach adopted in these sketches. Here he also included splashes of color to indicate flower plantings, not dissimilar to the suggestions of William Robinson. Crowning one of the sketches is a large octagonal pergola structure, no doubt intended to provide an architectural focal point to the whole composition. In no other project of this period did he allow himself such freedom in the distribution of planting, with almost no formal reference to the architectural garden, to the point that it is difficult to distinguish this proposal from a park in the older "landscape" style. These sketches show that Migge was capable of designing in different modes; he undoubtedly would have been able to explain the subtle perceptual effects that would distinguish his completed project from older landscape designs. In any case, he did not begin to design in this way again until the early 1930s, and the Hamburg sketches were only privately shown and quickly forgotten.

Later Interpretations of *Garden Culture*: Continuations and Qualifications

While Migge's *Garden Culture* didn't have quite the revolutionary effect that he was hoping for, it generally received very positive reviews in the professional press in Germany. It was a book with somewhat unusual viewpoints for the field, even at the time. One reviewer noted, positively, that "Migge is a restless spirit, who allows neither himself nor us any rest," further observing that he was "one of the very few people who have understood that the great economic transitions of capitalist production also demand artistic expression."[124] The historian Marie-Luise Gothein ended her magisterial two-volume work on garden history throughout the ages with a perspective drawing of Migge's Rheinhardt garden, showing her high regard for his work.[125] Migge's post–World War I writings on self-sufficient gardens and settlements probably had a broader readership than *Garden Culture*, and have certainly been discussed more since. Yet there have been a few significant later interpretations of *Garden Culture* that provide evidence of the book's perceived importance.

Within the English-language professional press of the time in Britain, there was virtually no mention of what was happening in German landscape design. But in the United States, *Garden Culture* was reviewed favorably in 1914 by the landscape architecture librarian at Harvard University, Theodora Kimball.[126] Harvard had traditionally been influenced by German academia, and the landscape architect Charles Eliot Jr., whose father, Charles William Eliot, was president of the university from 1869 to 1909, had written extensively about German landscapes, especially the work of Pückler-Muskau.[127] Later, in 1917, Kimball went on to write a treatise together with Henry Hubbard titled *An Introduction to the Study of Landscape Design*.[128] This was arguably the first treatise written by Americans attempting to systematically define the profession in their own country, while carefully considering the international context. The authors spent considerable time discussing the German contribution, described as a new kind of "formalism." For them, this new form of garden design was specifically German, even though it was initially inspired by English ideas. They admired "the adaptation of the different parts of the scheme each to its use," and "the arrangement of these parts for economic efficiency," especially in smaller, private gardens. But they believed that these were only "consistent and beautiful" when the designer was "blessed by nature with a sensitiveness to beauty of form." They expressed further reservations about the German work: "Especially in parks where the form of the design has been made to depend on its obvious and economic functions, with no consideration for the lightness of touch that comes from a certain waywardness, the result seems, esthetically, uninspired."[129]

Kimball and Hubbard continued by noting that *Garden Culture* was an "advanced exemplar" of the "propagandist literature" describing "this German formalism." Their general remarks on contemporary German gardens seem to be based on their reading of Migge, primarily in a positive sense, but with further cautionary statements: "What is to be the future of the *Gartenkultur* and *Gartentypen* which he so earnestly recommends for Germany, and apparently also for the rest of the world, will depend, however, on their fitness in use, on their adaptability to the actual needs of men."[130] While they discussed Migge at length, they did not provide any illustrations of his work, suggesting that they were more accepting of his overall logic than his actual designs.

Despite the reservations of Kimball and Hubbard, the chapter titled "Types of Landscape Designs" appears to be based on similar discussions in *Garden Culture*. The inclusive list of types that they give is very similar, with the addition of a few American types, such as the "country club" and "country hotel." In the opening of the chapter, they explain their own understanding of "the type": "Each typical well-defined use, however, has its more or less characteristic effect on the composition, no matter what the other circumstances may be. The result of this fact is that the ordinary work of the landscape architect falls into classes most readily according to use, and where this use is a common one,—one in which the habits of men are much alike,—the designs which serve this use will have much in common and may well be discussed together."[131] Kimball and Hubbard defined the type as an identifiable design solution based on functions established by social custom. They were not interested in universal solutions as such, nor did they take up Migge's cry for gardens for the masses, except in an implicit sense. They showed no great concern for the underlying forces of capitalism, industrialization, or metropolitan growth, but they accepted Migge's emphasis on organization and systematic thinking. As professional landscape architects, they appreciated some qualities of Migge's designs, but were concerned that such design principles not be too rigidly applied, potentially excluding "the lightness of touch that comes from a certain waywardness."[132] From a historical perspective, this formulation of the type may be seen as following a transcultural line of development, from the English garden designer Thomas Mawson to the German architect Hermann Muthesius, and through Migge to the Americans Hubbard and Kimball, who arguably, in turn, influenced other landscape architects in the United States.

There does not appear to have been very much discussion of *Garden Culture* in either English or German for several decades. Probably the next important study of *Garden Culture* is found in *The Park and the Town* (1966), by the English planner and historian George Chadwick.[133] Chadwick was a product of modernist planning and landscape-design thinking in post–World War II Britain, and therefore a strong believer in the necessity of public green space for the social well-being of the public.[134] As to be expected, as a modernist himself, he saw Migge as a true pioneer: "With the work of Leberecht Migge and others, though, just prior to World War I, German garden design really did shake off the deadening influence of styles and produce work which probably ranks among the first essays in truly modern landscape."[135] By contrast, Chadwick observed that the Märchenbrunnen (Fairytale Fountain) design by Migge's contemporary Ludwig Hofmann in Berlin had not achieved this newness of expression because it still relied on historicist form and ornamentation. Written in the 1960s, this was indeed a compliment. Chadwick raised the problem of the naturalistic landscape park decades after the original debates, here in a more international context: "Migge's own designs, and those of his associates, show a distinct change in attitude from the Olmstedian concept of recreation spaces set in a large landscape in the Repton manner, to that of the formation of the whole design around a series of spaces whose primary requirements were those demanded by a particular recreational activity: function now began to create form, instead of being accommodated within a form which was preconceived."[136] Chadwick, as a modernist, read into Migge's park design work the maxim "form follows function," which indeed would have been fully in keeping with Migge's own stated intentions in *Garden Culture*. Chadwick also implied that Migge's functional approach was superior to the older, more accepted "Olmstedian" ideal of the urban pastoral. Chadwick was referring specifically to *Garden Culture* when he wrote, "Migge's published designs of public work, although tiny reproductions, are worth study for their development of the basically simple means of expression adopted." Thus, Migge's 1913 book, even though published before the high-modern period of the 1920s, represented a new, modernist form of landscape expression that still resonated with the design mentality of the 1960s.

When Migge's work was discussed in Germany after World War II, it was usually in reference to his ecological design ideas, not to the principles expressed in

Garden Culture. One important exception was the writing of the so-called Kassel School from the late 1970s through the 1980s.[137] The members of the Kassel School were landscape architects of a generation that was criticizing the high modernism of the 1950s and 1960s through a socially aware agenda. As they themselves wrote, they were not professional historians. Rather, as designers, they were interested in Migge for the lasting relevance of the concepts he put forth in 1913 in *Garden Culture*.[138] Their overall understanding of "garden culture" is summarized in the following statement: "Migge's apparently banal statement, 'Making a good garden is in principle one of the most simple and matter-of-fact things: *one must organize it*,' doesn't just present old gardens as precedents, rather it is valid for villages, settlements, cities, and the organization of their open spaces. Migge's call for 'functionality, simplicity, and economic quality,' is based on the rich store of experience of the history of many centuries of the use of the house, the yard, the garden, the street, the path, and the square."[139]

Here "experience" is the key word, for the Kassel School writers were thinking more on phenomenological lines about traditional spatial entities, rather than of a kind of biological evolution of history based on abstract form types, as Migge had done. To put it another way, they were arguing for a consideration of traditions of use and identification of place, rather than purely formal solutions derived from "functions." Unlike Chadwick, the Kassel School writers contrasted Migge to high-modernist landscape design, which they associated with the group around Le Corbusier: "The ideological highpoint was reached with the 'Athens Charter,' by which municipal gardeners unified the pseudo-romantic landscape with 'integrated' buildings designed as *staffage*."[140] They appear to be referring to the kind of "pastoral" open spaces around modernist high-rise housing, so familiar around the world. For them, these landscapes were not functional in the sense of contributing to daily domestic life, but were merely "pseudo-romantic" revivals of the nineteenth-century landscape park, with the modernist buildings acting almost as eye-catchers. Instead, the Kassel School proposed that spaces be organized to accommodate the greatest range of active uses, much as Migge's green outdoor rooms. They understood the traditional unity of house and yard as "productive and reproductive" space, functioning as an integral part of family life.[141] While this reading of *Garden Culture* was very much a reflection of their own agenda, it nonetheless reveals the multitude of possible interpretations. They claimed to have had no interest in the historical context, and thus excused themselves from acknowledging the top-down, paternalistic attitude of social reformers in Migge's era. Instead, they took Migge's arguments for social inclusion at face value, which they interpreted as admonitions to the design professions that had not lost their worth.

In all of these examples, the theme from the earlier garden-reform era of the alleged duality between the aestheticized landscape garden and the functional, well-organized green space plays a background role, however defined. Hubbard

EUGEN DIEDERICHS VERLAG IN JENA

Im Oktober 1913 erscheint
mit Unterstützung der Kgl. Bauakademie in Charlottenburg

M. L. GOTHEIN/GESCHICHTE DER GARTENKUNST. 2 Bde. mit ca. 600 Illustrationen à ca. M 15.—

Inhalt: Ägypten / Westasien im Altertum / Griechenland / Das römische Reich / Byzanz und die Länder des Islam / Das mittelalterliche Abendland / Italien im Zeitalter der Renaissance Spanien und Portugal im Zeitalter der Renaissance / Frankreich im Zeitalter der Renaissance / England im Zeitalter der Renaissance / Das Zeitalter der Renaissance in Deutschland und in der Niederlande / Das Zeitalter Ludwigs XIV. / Die Ausbreitung des französischen Gartens in Europa, England, Deutschland, Schweden, Dänemark, Rußland, Italien, Spanien, Holland / China und Japan / Der englische Landschaftsgarten / Hauptströmungen der Gartengestaltung im 19. Jahrhundert bis zur Gegenwart.

Dieses nicht nur gelehrte, sondern auch künstlerisch anschaulich geschriebene Werk erfüllt für jeden gärtnerischen Fachmann und Gartenliebhaber einen lang gehegten Wunsch. Es ist die erste vollständige, auf langjähriger Beschäftigung beruhende Orientierung über den inneren Zusammenhang des Wechsels der Stilformen des Gartens. Es basiert direkt auf den literarischen Quellen und rückt viele Einzelheiten zum ersten Male in eine fruchtbare Beleuchtung. Ja, es bringt geradezu neue Entdeckungen, z. B., aus welchen Einflüssen der englische Gartenstil entstand. Auf die Sammlung des Illustrationsmaterials wurde besonders viel Sorgfalt verwandt; es sind alle historisch wichtigen Gärten Europas neben vielen interessanten Einzelabbildungen vertreten.

FRITZ ENCKE/DER HAUSGARTEN. Mit 115 Illustr. br. M 5.—, Lwd. geb. M 6.—

Der Schrebergärtner: Unsere Gartenbauliteratur, soweit sie sich mit der Gartenkunst befaßt, wird durch dieses hervorragende Werk um ein wertvolles bereichert. Zeigt Encke in seinem Buch auch keinen Hausgarten, in dem man praktischen Gartenbau treiben und Obst und Gemüse bauen kann, wie es von einem eigentlichen Hausgarten doch wohl von den meisten verlangt werden wird, so führt er uns mehr den ästhetisch gehaltenen Garten vor, dessen Ausbildung er auch — nach den Abbildungen des Buches zu urteilen — meisterhaft versteht. Dieses Vertiefen in die Eigenheiten des jeweiligen Grundstücks verrät den Künstler, der ganz in seinem Berufe aufgeht. Und mit welchen geringen Mitteln weiß der Verf., der übrigens Gartendirektor der Stadt Köln ist, hervorragend schöne und natürliche Szenerien zu schaffen; ein dünner Wasserfaden, der Ablauf eines am Haus befindlichen, in seiner Schlichtheit herrlichen Brunnens genügt, um den ganzen Garten zu beleben. Er tritt als kleiner reizender Wasserfall auf und macht als solcher einen derartig natürlichen Eindruck, daß man meint, er habe sich dort schon vor Anlage des Gartens befunden, und das ist unbedingt Gartenkunst. Besondere Beachtung beanspruchen die beigefügten Gartenpläne; manchem, der an Schema F gewöhnt ist, werden sie allerdings unverständlich sein; doch so sollen Gartenpläne sein; sie sollen, sie müssen auf das Gelände passen. Auf Vorrat lassen sich Gartenpläne nicht anfertigen, deshalb hat dieses Buch mit seinen 3 Gartenplänen auch mehr wert, als manches, das deren 50—100 enthält, die gar zu oft Ursache geworden sind, die Gartenkunst in Mißkredit zu bringen. Dem Buch ist unbedingt weiteste Verbreitung zu wünschen.

figure 11

Original book advertisements included at the back of *Garden Culture* for the forthcoming two-volume work by Marie Luisa Gothein, *Geschichte der Gartenkunst* (*History of Garden Design*), still one of the most authoritative works on the subject; and for Fritz Encke's *Der Hausgarten* (*The House Garden*), previously published in the same series in which *Garden Culture* appeared. Eugen Diederichs was a prominent publisher committed to cultural reform in Germany and a member of the German Werkbund.

and Kimball openly rejected this alleged duality, instead arguing that there should be room for the unplanned, for moments of chance—and presumably for something resembling the naturalistic landscape mode—together in a unified whole with more architectural and functional elements. The concepts of organization and the type were important for these two landscape writers, but not to the exclusion of aesthetic caprice. Chadwick took a more definite position; believing in the modernist social agenda of the post–World War II era, he rejected the Olmstedian pastoral in favor of the functionalist approach to green-space planning. Finally, focusing on a different strain of modernism from that of Chadwick, the Kassel School critics equated the generic green spaces of the modernist pastoral with the old naturalistic landscape mode, finding in them another kind of empty aestheticism, perhaps only updated. For this group, the modernist pastoral limited human activity rather than supported it, especially when applied to domestic life within social housing. They were quick to point out, however, that they were not trying to advance Migge's actual design solutions, which they found somewhat flawed (following their own line of thinking). The writers of the Kassel School proposed that the appearance of a designed landscape should be dictated not by style or form but by use, the latter based on the richness of human experience rather than on prescribed design formulas.

Perhaps *Garden Culture* will continue to inspire new arguments and discussions, especially now that it is available in English in its entirety for the first time. As a historical document, it gives insight into the philosophies that shaped design thinking during the period, and how these later informed the high modernism of the 1920s. On another level, Migge's designs could be read as good, solid examples of functional green-space planning, which, in the case of the private gardens at least, reached a high degree of spatial complexity and variety in the planting compositions. On the other hand, for the Kassel School, the most important aspects of Migge's thinking were those based on social and economic principles, reinforced by his general arguments for organization and functionality. Given that today many writers and designers see landscape "design" as being driven by resources and infrastructure in the metropolitan and regional contexts, Migge's examination of underlying forces could take on new relevance.[142] Yet there is clearly a disjunction in *Garden Culture* between Migge's analyses of metropolitan, mass culture, and his actual design solutions, which are rather prescribed in terms of their forms. It could be argued, though, that a looser, more generous reading of the actual designs could be made in light of his conceptual arguments. In any case, whether or not historical background or Migge's formal preferences are taken into account, later reception of *Garden Culture* suggests that the depth and complexity of his arguments continue to allow for a variety of new insights and perceptions.

Notes to the Translator's Introduction

Citations from this translation will be given in the introductory chapter as follows: GK stands for the original page number, and GC for the page number in the English translation.

1 GK ii, GC 48.

2 GK 157, GC 197.

3 See Hermann Muthesius, "Wo stehen wir?" in *Jahrbuch Deutschen Werkbundes 1912: Die Durchgeistigung der Deutschen Arbeit; Wege und Ziele in Zusammenhang von Industrie/ Handwerk und Kunst* (Jena: Euegen Diederichs, 1912), 16. For general biographical information on Muthesius, see Barbara Volkman and Sonja Günther, eds., *Hermann Muthesius: 1861–1927* (Berlin: Akademie der Künste, 1977). The most complete study of Muthesius's role in garden reform and of his own garden designs is Uwe Schneider, *Hermann Muthesius und die Reformdiskussion in der Gartenarchitektur des 20. Jahrhunderts* (Worms: Wernersche Verlagsgesellschaft, 2000).

4 See Fritz Encke, *Der Hausgarten* (Jena: Diederichs, 1907); Willy Lange, *Gartengestaltung der Neuzeit* (Leipzig: J. J. Weber, 1907); Camillo Schneider, *Landschaftlicher Gartengestaltung* (Leipzig: Scholtze, 1907); Reginald Blomfield, *The Formal Garden in England* (London: Macmillan, 1892); Thomas H. Mawson, *The Art and Craft of Garden Making* (London: Batsford, 1901); William Robinson, *The Wild Garden* (London: J. Murray, 1870); André Vera, *Le nouveau jardin* (Paris: Émile-Paul, 1912); and Theodora Kimball and Henry Hubbard, *An Introduction to the Study of Landscape Design* (New York: Macmillan, 1917).

5 Probably the most strongly nationalist among contemporary German garden treatises is by Hampel, as he completely rejects foreign influences. Hampel belonged to the older generation of German garden designers, who generally opposed reform: Carl Hampel, *Die deutsche Gartenkunst* (Leipzig: Hugo Voight, 1911), 3–4. In this passage, Hampel asserts that the Germans are not as sporting as the English, instead needing gardens only for passive contemplation.

6 Thomas H. Mawson, *Civic Art* (London: Batsford, 1911); J. C. N. Forestier, *Grandes villes et systèmes de parcs* (Paris: Hachette, 1906); and Hugo Koch, *Gartenkunst im Städtebau* (Berlin: Wasmuth, 1914).

7 See Christiane Crasemann Collins, *Werner Hegemann and the Search for Universal Urbanism* (New York: W. W. Norton, 2005).

8 Werner Hegemann, *Ein Parkbuch: Amerikanische Parkanlagen* (Berlin: Wasmuth, 1911).

9 At this time, Migge was still working for Jakob Ochs in Hamburg.

10 Paul Schultze-Naumburg, *Die Kultur des weiblichen Körpers als Grundlage der Frauenkleidung* (Leipzig: Diederichs, 1903).

11 Further details and all references in this biographical sketch are in David H. Haney, *When Modern Was Green: Life and Work of Landscape Architect Leberecht Migge* (Abingdon, UK: Routledge, 2010).

12 Leberecht Migge, *Der Hamburger Stadtpark und die Neuzeit* (Hamburg: Brücker, 1909).

13 Leberecht Migge, *Jedermann Selbstversorger!* (Jena: Diederichs, 1919).

14 I use "gardenesque" as a general term to describe changes taking place in the early nineteenth-century garden, on both the villa and the landscape scales. Loudon indeed coined the term "gardenesque," and provided his own explanations of it, but he was also capturing the spirit of the period; as with many artistic movements, such changes were already in the air at the time. One garden designer who epitomized this approach in the mid-nineteenth century was Joseph Paxton. See J. C. Loudon, *The Suburban Gardener and Villa Companion* (London, 1838); and Melanie Simo, *Loudon and the Landscape: From Country Seat to Metropolis* (New Haven: Yale University Press, 1988). Another historian has previously taken a similar position to my own; see George F. Chadwick, *The Park and the Town* (London: The Architectural Press, 1966), 251: "As in England towards the close of the [nineteenth] century, there was in Germany, too, a reaction against the Gardenesque which had prevailed almost without question for the greater part of the century."

15 It should be noted that the gardener Joseph Paxton designed the exhibition building (the "Crystal Palace") as well as the extensive grounds for the building when it was relocated to Sydenham (London) in 1854. I would argue that Paxton's method of garden design was very similar to the gardenesque as described by Loudon. See Kate Colquhoun, *A Thing in Disguise: The Visionary Life of Joseph Paxton* (London: Fourth Estate, 2003).

16 Mark Girouard, *Sweetness and Light: The Queen Anne Movement, 1860–1900* (New Haven: Yale University Press, 1990), 152–59.

17 Robinson, *Wild Garden*; see also the comprehensive biography by Richard Bisgrove, *William Robinson: The Wild Gardener* (London: Frances Lincoln, 2008), and the critical analysis (especially in relation to Germany) by Anne L. Helmreich, "Re-presenting Nature: Ideology, Art, and Science in William Robinson's 'Wild Garden,'" in *Nature and Ideology: Natural Garden Design in the Twentieth Century*, ed. Joachim Wolschke-Bulmahn (Washington, D.C.: Dumbarton Oaks Research Library and Collection, 1997), 81–111.

18 William Robinson, *The English Flower Garden*, 6th ed. (London: J. Murray, 1898), 6.

19 Blomfield, *Formal Garden in England.*

20 How exactly these were to be defined was the source of many theoretical debates, such as that between Migge and Willy Lange.

21 "pretzel paths," *Brezelwege*. Friedrich Avenarius, "Piepenbrincks im Garten," *Der Kunstwart* (1899): 206–7.

22 Again, I am using "gardenesque" as a general term; the English garden for the older generation of German garden designers, not the reformers, would have been something like the gardenesque rather than the newer Arts and Crafts or wild gardens. German reform writers did not necessarily use the term "gardenesque," as they generally referred instead to the "Lenné-Meyer" School. On the latter term, see Schneider, *Hermann Muthesius und die Reformdiskussion in der Gartenarchitektur des 20. Jahrhunderts*, 230–31. On the history of the professionalization of garden design during this period, see Gert Gröning and Joachim Wolschke-Bulmahn, *DGGL, Deutsche Gesellschaft für Gartenkunst und Landschaftspflege e. V.: 1887–1987; Ein Rückblick auf 100 jahre DGGL* (Berlin: Boskett, 1987).

23 The original article was later republished as Alfred Lichtwark, *Makartbouquet und Blumenstrauß* (Berlin: Bruno Cassirer, 1905).

24 Gebhart suggested that Lichtwark's interest in the "original" garden was derived from his reading of Gottfried Semper, who emphasized a return to architectural origins. Julius Gebhart, *Alfred Lichtwark* (Hamburg: Hoffmann und Campe, 1947), 41.

25 Alfred Lichtwark, *Park- und Gartenstudien* (Berlin: Bruno Cassirer, 1905).

26 A third project, the Cramer House in Berlin Zehlendorf (1910), may have been an additional collaboration, based on the fact that the Berlin branch of the Ochs firm provided the planting materials, but this has not been substantiated. See the client list in Brita von Husen

and Heino Grunert, "Der Gartenarchitekt Jakob Ochs (1871–1927) und sein Atelier," *Die Gartenkunst* 1 (1999): 54–80.

27 The section on garden design is in Hermann Muthesius, *Das englische Haus*, vol. 2, *Bedingungen, Anlage, gärtnerische Umgebung, Aufbau und gesundheitliche Einrichtungen des englischen Hauses* (Berlin: Wasmuth, 1910). The most recent English translation is Hermann Muthesius, *The English House*, trans. Janet Seligman and Stewart Spencer (London: Frances Lincoln, 2007). See also Schneider, *Hermann Muthesius und die Reformdiskussion in der Gartenarchitektur des 20. Jahrhunderts*.

28 Mawson, *Art and Craft of Garden Making*, 16; and Muthesius, *Das englische Haus*, 2:82.

29 In the English translation by Seligman and Spencer, the word *regelmässig* ("regular") is consistently translated as "formal," which may sound more normal to English speakers but which in fact obscures the meaning Muthesius was trying to convey. See Muthesius, *English House*, 96.

30 On Jekyll, see Jane Brown, *Gardens of a Golden Afternoon* (London: Penguin, 1994).

31 Hermann Muthesius, *Landhaus und Garten* (Munich: Bruckmann, 1907). Although the majority of Muthesius's *Landhaus* projects were large suburban villas with grounds, in some cases he designed houses in the country, some of which were very extensive. See the published catalogue of his significant *Landhaus* projects: Hermann Muthesius, *Landhäuser* (Munich: Bruckmann, 1912).

32 I have generally used *Landhaus* rather than "country house" in my translation here, because Muthesius used this to denote a new form, as opposed to the historicist "villa." But I have used "country houses" for the plural form (*Landhäuser*), because the German plural form is unfamiliar to most English-speaking readers.

33 Muthesius, *Landhaus und Garten*, xxii.

34 A chicken house was often elevated and provided with a ramp or ladder for chickens (*Hühnersteig*), as a means of protecting them from marauders. Muthesius, *Landhaus und Garten*, xxii.

35 Ibid., xxvi.

36 See note 29.

37 Muthesius, *Landhaus und Garten*, xxii.

38 GK 88, GC 129.

39 See, for example, Lichtwark's comment on Robinson's influence in Alfred Lichtwark, *Blumenkultus, Wilde Blumen* (Berlin: Cassirer, 1907), 74. Lichtwark cited Jekyll on many occasions.

40 Hans-Joachim Hubrich, *Hermann Muthesius: Die Schriften* (Berlin: Gebr. Mann, 1981), 151–52.

41 GK 92, GC 135.

42 GK 92, GC 135.

43 GK 96–97, GC 139.

44 See Plates 16 and 22.

45 GK 111, GC 151.

46 GK 111, GC 152.

47 GK 120, GC 158.

48 GK 120, GC 158.

49 GK 95, GC 139.

50 Lange, *Gartengestaltung der Neuzeit*.

51 Lange used numerous terms for "plant communities," such as *Pflanzengenossenschaftsbilder* (lit. "plant cooperative forms") and more literally, *Pflanzen-Lebensgemeinschaften* (Lange, *Gartengestaltung der Neuzeit,* 170). In the third edition of *Gartengestaltung der Neuzeit,* Lange discussed ecology and biology as a basis for "natural" garden design (Willy Lange, *Gartengestaltung der Neuzeit* [Leipzig: J. J. Weber, 1912], 19, 221–22). Lange most often used the term "natural" garden, but he made clear that it was based on ecological principles: "Die Lehre von den Lebensgemeinschaften ist der Naturkunde als ökologische Biologie heute geläufig. Es braucht aber nicht Jahrzehnte zu dauern, bis sie für die Ästthetik [misprint: Ädie sthetik] nutzbar gemacht wird" (ibid., 221–22). In my introduction, I use the term "ecological," in addition to "natural," to emphasize the intended scientific significance.

52 GK 144, GC 183–84.

53 See Olaf Breidbach, *Visions of Nature: The Art and Science of Ernst Haeckel* (Munich: Prestel, 2006).

54 Ernst Haeckel, *Die Natur als Künstlerin* (Berlin: Vita, 1913).

55 GK 92, GC 135.

56 GK 66, GC 110.

57 GK 66, GC 110.

58 The most thorough and comprehensive study of the Werkbund in English, on which much of the discussion here is based, is Frederic J. Schwartz, *The Werkbund: Design Theory and Mass Culture before the First World War* (New Haven: Yale University Press, 1996).

59 What was meant by the term "spiritual" (*geistig*) in this era is complex and elusive; the spiritual was read by some as the life force in living beings, drawing upon the writings of the biologist Ernst Haeckel, and related conceptually to vitalism. It was thus not a reference to organized religion as such. In addition, the German word *Geist* can mean either "spirit" or "intellect"; here, I have followed other English-speaking historians on the Werkbund, especially Frederic J. Schwartz, and have myself used derivates of the term "spiritual" in most cases.

60 Friedrich Naumann, *Kunst und Volkswirtschaft* (Berlin: Werkbund, 1912), 8.

61 The German equivalent terms "landscape architecture" (*Landschaftsarchitektur*) and "landscape architect" (*Landschaftsarchitekt*) were not in use until after World War II.

62 GK 91, GC 133.

63 GK 139, GC 179.

64 GK 141, GC 181–82.

65 GK 143, GC 183.

66 GK i, GC 47.

67 Schwartz discusses the importance of the type for Muthesius and the Werkbund throughout *The Werkbund: Design Theory and Mass Culture before the First World War.*

68 Hubrich, *Hermann Muthesius,* 151–52.

69 Werner Oechslin, "Sempers Kosmos," in *Gottfried Semper 1803–1879: Architektur und Wissenschaft,* ed. Winfried Nerdinger and Werner Oechslin (Munich: Prestel, 2003).

70 Tea kettle images also reproduced in Schwartz, *Werkbund,* 130–31.

71 GK 145, GC 184.

72 GK 145, GC 184.

73 GK 146, GC 186.

74 GK 149, GC 189.

75 GK 157, GC 197.

76 GK 157, GC 197.

77 For a comprehensive analysis of Nietzsche's aesthetic theories, see Theodor Meyer, *Nietzsches Kunstaufassung und Lebensbegriff* (Tübingen: Fancke, 1991). For his "life philosophy" (*Lebensphilosophie*), see pp. 121–35, 597–98.

78 For a discussion of Muthesius's use of Nietzschean terms, see Hubrich, *Hermann Muthesius*, 74.

79 For an analysis of Kantian theory in relation to late eighteenth-century German garden aesthetics, see Michael G. Lee, *The German "Mittelweg": Garden Theory and Philosophy in the Time of Kant* (London: Routledge, 2007), 63.

80 On the critique of civilization, and on Tönnies in relation to art and design, see Richard Hamann and Jost Hermand, *Stilkunst um 1900* (Munich: Nymphenburger Verlagshandlung, 1975), 102–6.

81 GK ii, GC 48.

82 GK ii, GC 48.

83 For a wide-ranging series of articles on life reform (*Lebensreform*), including the influence of Nietzsche, see Kai Buchholz et al., eds., *Die Lebensreform: Entwürfe zur Neugestaltung von Leben und Kunst um 1900*, 2 vols. (Darmstadt: Institut Mathildenhöhe, 2001).

84 GK 7, 153, GC 57–58.

85 The concept of *Kunstwollen* is introduced in Alois Riegl, *Stilfragen* (Berlin: Siemens, 1893). For an analysis of the influence of Riegl's thinking on Muthesius and others in the Werkbund, see Schwartz, *Werkbund*, 21.

86 As mentioned in note 59 above, the "spiritual" is a difficult term for contemporary readers to comprehend. For a thorough introductory discussion of what was meant by the spiritual in this context, see Schwartz, *Werkbund*.

87 Wilhelm Wundt, *Grundriss der Psychologie*, 10th ed. (Leipzig: Engelmann, 1911), 177–78, 200–1, 204. On Muthesius's frequent use of the term "rhythm," see Hubrich, *Hermann Muthesius*, 74–75.

88 GK 21, GC 71.

89 GK 66, GC 110.

90 GK 150, GC 189.

91 Karl Bücher, *Arbeit und Rhythmus*, 4th ed. (Leipzig: Teubner, 1909), 24.

92 C. C. L. Hirschfield, *Theory of Garden Art*, ed. and trans. Linda B. Parshall (Philadelphia: University of Pennsylvania Press, 2001).

93 The German term *Großstadt* may be translated variously as "big city," "great city," or "metropolis." I have varied the usage as seemed appropriate in context.

94 GK 7, GC 58.

95 August Endell, *Die Schönheit der großen Stadt* (Stuttgart: Strecker and Schröder, 1908).

96 Karl Scheffler, *Die Architektur der Gross-Stadt* (Berlin: Cassirer, 1913), 6.

97 GK 22, GC 71.

98 GK 36, GC 83.

99 Migge appears to be the author of a section on the work of the Ochs firm, included in the original publication.

100 Hegemann, *Ein Parkbuch*.

101 See Plate 5 and Appendix 7.

102 GK 36, GC 83.

103 GK 7, GC 58.

104 GK 9, GC 59. *Bürgergarten* has been translated as "citizen's garden" in the context of the eighteenth century (e.g., Parshall's translation of *Theory of Garden Art*); however, in the early

twentieth century the social classes were much more differentiated than in the late eighteenth century. Therefore, I have used the term "bourgeois" to signify the comfortable, solid middle classes, which is effectively what Migge meant.

105 Günther Katsch and Gerhild Schwendler, eds., *Der Schreberverein und der Verband Leipziger Schrebervereine 1864–1922* (Leipzig: Deutsches Kleingärtnermuseum, 2004).

106 GK 18, GC 68.

107 GK 21, GC 71.

108 George R. Collins and Christiane Crasemann Collins, *Camillo Sitte: The Birth of Modern City Planning* (New York: Rizzoli, 1986). See also Camillo Sitte, *Der Städtebau nach seinen künstlerischen Grundsätzen*, 4th ed. (Vienna: K. Graeser, 1909), which contains an additional section on *Großstadtgrün* ("metropolitan green space").

109 GK 28–29, GC 77–78.

110 GK 25, GC 75.

111 GK 28, GC 77.

112 GK 35, GC 83.

113 Dohna-Poninska discussed the importance of "open space" (*im Freien*) and described the social necessity of outer parks much as Migge did four decades later. Adelheid (Arminius) Dohna-Poninska, *Die Großstadt in ihrer Wohnungsnoth und die Grundlagen einer durchgreifenden Abhilfe* (Leipzig: Duncker and Humblot, 1874), 137, 148, 239–40.

114 Richard Baumeister, *Stadt-Erweiterungen* (Berlin: Ernst and Korn, 1876), 185–95.

115 See, for example, Alfred Leutscher, *Epping Forest: Its History and Wildlife* (Newton Abbot: David and Charles, 1974).

116 Clemens Alexander Wimmer, *Parks und Gärten in Berlin und Potsdam*, 3rd ed. (Berlin: Nicolaische Verlagsbuchhandlung, 1989), 47–51.

117 For a good general discussion of the influence of the garden-city idea in German-speaking countries, see Franziska Bollerey and Gerhard Fehl, eds., *Im Grünen wohnen—im Blauen planen: Ein Lesebuch zur Gartenstadt mit Beiträgen und Zeitdokumenten* (Hamburg: Christians, 1990). On the German Garden City Association specifically, see Kristiana Hartmann, *Deutsche Gartenstadtbewegung: Kulturpolitik und Gesellschaftsreform* (Munich: Heinz Moos, 1976).

118 Migge reports on his summer trip to England with the German Garden City Association in "Mehr Ökonomie," *Die Gartenstadt* 10 (1910): 109–13.

119 See Migge, *Jedermann Selbstversorger!* I have translated this as *Everyman Self-Sufficient!*; this is not an exact translation, but it captures the slogan-like spirit of the title.

120 For a more complete discussion of these parks (and additional sources), see Haney, *When Modern Was Green*, 55–58, 72–85.

121 On the possible influence of Prince Hermann von Pückler-Muskau, see David H. Haney, "Bringing the Americanized Pückler Back to Germany: Charles Eliot and the German Park Reform Movement," in "Pückler in America," supplement 4, *Bulletin of the German Historical Institute* (2007): 89–110.

122 I am the first to have identified these sketches, which were never publicly discussed or previously published. (The original sketches are in the collection of the heirs of Leberecht Migge.)

123 Fritz Schumacher, *Ein Volkspark* (Munich: Callwey, 1928).

124 Fritz Hellwag, review of *Die Gartenkultur des 20. Jahrhunderts*, by Leberecht Migge, *Kunstgewerbeblatt* 25, no. 5 (1914): 99.

125 Marie-Luise Gothein, *Geschichte der Gartenkunst*, vol. 2 (Jena: Diederichs, 1914).

126 Theodora Kimball, review of *Die Gartenkultur des 20. Jahrhunderts,* by Leberecht Migge, *Landscape Architecture* 1 (1914): 54.

127 On the landscape architect Charles Eliot Jr. and his travels through Germany, see Charles William Eliot, *Charles Eliot, Landscape Architect* (Boston: Houghton Mifflin, 1902), reprinted with an introduction by Keith Morgan (Amherst: University of Massachusetts Press, 1999).

128 Theodora Kimball and Henry Hubbard, *An Introduction to the Study of Landscape Design* (New York: Macmillan, 1917).

129 Ibid., 52–53.

130 Ibid., 54.

131 Ibid., 231–32.

132 Hubbard and Kimball make a positive mention of Migge's design for the Mariannenpark in Leipzig, but the fact that they provide no illustration of it suggests a certain ambivalence. Hubbard and Kimball, *Introduction to the Study of Landscape Design,* 226.

133 Chadwick, *Park and the Town.* See note 14.

134 George Chadwick received his master's degree in town planning in 1950 at the University of Manchester with a thesis titled "The Character of Open Spaces in the Modern Town Plan." His PhD dissertation on Joseph Paxton was awarded in 1959. His PhD supervisor was the architecture professor Clifford Holliday, who had spent significant time in Palestine in the 1930s. Chadwick was unusual in Britain for the international scope of his knowledge on park design. (Sources: Paul Brenikov and Peter Willis, Newcastle University; John Archer, University of Manchester; and Manchester University Archives)

135 Chadwick, *Park and the Town,* 251, 254.

136 Ibid., 254–55.

137 Helmut Böse-Vetter, ed., *Notizbuch 10 der Kasseler Schule: Nachlese Freiraumplanung* (Kassel: Kassel University, Arbeitsgemeinschaft Freiraum und Vegetation, 1989).

138 Ibid., 16.

139 Ibid., 16.

140 Ibid., 24.

141 In the writings of the Kassel School, "production" and "reproduction" are taken from standard Marxist terminology.

142 I am thinking here, for example, of so-called landscape urbanism. See Charles Waldheim, ed., *The Landscape Urbanism Reader* (New York: Princeton Architectural Press, 2006).

LEBERECHT MIGGE
DIE GARTENKULTUR DES 20. JAHRHUNDERTS

MIT 74 JLLUSTRATIONEN UND 23 TAFELN

VERLEGT BEI EUGEN DIEDERICHS IN JENA / 1913

FOREWORD

Building, clothing, and utensils are what we usually think of when the subject is culture. In the final and essential sense though, all of these crude things are only of relative importance for the cultural maturity of our race. These allow for the improvement of our living conditions, nothing more. Only as *obstructions* to deeper levels may they take on importance. Then we indeed know or clearly sense: by their own nature, humans will never cease to tinker with the physical environment, never stop until a form appears that gives lasting satisfaction. For a given period of time, things are so perfect in themselves and belong so harmoniously together, and are so reliable, beautiful, and familiar that our highest energies are no longer required. One then thinks that human products are finished; they have style. Style is simplification and unity, it is rest. Once a nation or generation has reached such a state of rest, a tendency toward the cultivation of *inner* values emerges. And similarly the individual person in his day-to-day life first recognizes the possibility of a true bodily and spiritual, that is, personal culture, when by some means he has solved the problems of forms that surround him.

It is not much different with the garden. It is an embarrassing error to think that luxurious gardens with beautiful forms inherently result in garden *culture*. They only comprise and fulfill this term to a limited extent, just as if beautiful houses or cities alone would guarantee good or happy occupants.

But the garden has something in itself above the other trappings of our existence. These may be great, practical, hygienic, and who knows what else. The garden offers this as well, but it offers something more, something rare and irreplaceable: *it is living*. Everything wakes and grows in it and is not dead like cold stone and hard timber. The constant intimate contact with living nature that garden life entails elates us, inclining us to all that is noble and great. The unshakable law underlying all growth reveals the beneficial power of order, of the far-seeing will to plan. The obviously ephemeral character of even that which is most powerful and striking is dramatically revealed in the essence of the garden, giving comfort to the tired and awaking brotherly feelings in the strong. As to the wonderful rhythm expressed in the structure and color of plants and flowers, it sharpens even the most basic sensibility, revealing the essence of beauty and perfection: Pray tell, is there anything among the good things of this world that provides so much instruction, stimulation, and opportunities for the refinement of humanness in people than the garden? Our garden full of fresh open air, full of flowers, of green, and full sun?

But in order to realize and enjoy the immeasurable gifts of the garden, first we must sweep aside the great rubbish and detritus that covers over the field of gardening, as scarcely any other. There one accomplishes nothing: the state of activity that we see here is only appearance. Among the informed, this complicated garden-web[1] that now enmeshes us, this zealous preaching, planning, and testing, only signifies a veil over the chaos that spreads underneath. How will our new gardens come about, from where will they come, and where will they go? These are questions that will affect everyone—sooner or later.

Since I am trying to address these in a direct manner, I must consider without exception all garden phenomena following the same underlying intellectual, social, and economic factors that give birth to these and everything else. This somewhat unusual method touches the most sensitive points of our communal life and other concerns. Does the dominant capitalist economic system really encompass its own ethical values and the means of elevating them? Should the great misunderstanding between quality and income, between beauty and technique, continue to dominate? Is it possible more than previously to organically integrate the creative energies of our people with the actual dominant economic forces, denoting their common area and objectively assessing their worth? And that means nothing more or less than transplanting something of the great conflict of our time between rhythm and muscle, between spirit and money, within the blooming fields of our garden, thus reminding artists, technicians, and merchants of their common origins and their essentially common goals, spurring them on to mutual agreement and action. Will this new German gardening be the first among our 101 professions that strives to create a communal, goal-oriented cultural project from out of a dry, complex technical field? And since all great areas of creative activity are similarly ailing, should this remain the only such attempt in Germany?

Yes, it would be only an attempt at first. But already at this preliminary stage it is desirable, nay, as seen from *all* sides, it is *necessary*, to bring the external driving mechanism of gardening into a definite system, to simplify and popularize it—this book was in fact written to demonstrate this. The garden must be *organized*. Its material, its creator, and its owner must each be able without effort to find and appreciate their correct place within the overall operation. It is meant in earnest when I say: we must *overcome* today's "garden" if we want to achieve garden culture.

This is necessary, leaving aside all refinements, even as a pure bodily life need. Today's individual person scarcely can do without garden living in the long term. This will become ever clearer. And, for our frighteningly destructive civilization, the garden will be a conserving counterweight and an inexhaustible source of new creative energies for the nations.

So for now as means to an end, may the understanding of gardens and gardening everywhere become the starting point, the way, or even the object of a new, more just, and more beautiful coexistence among humans.

Hamburg-Blankenese, Spring 1913
LEBERECHT MIGGE

PLATE 1

THE ILLUSTRATIONS

In the beginning, this was to be a rather large picture book, as demanded today by the children of our age. And in fact there would have been hardly any obstacles to this: there was a sheer mass of stimulating material available. But we realized early on that among other significant opposing factors, if this work had been so designed, it would have become much too expensive to be accessible for those to whom the content is directed. So the project has gone forward with the illustrations being limited to what is necessary and in fact new.

The following materials were graciously provided by Messrs. Dipl.-Ing. *Martin Wagner* (p. 12 [62][2]), Head of the Rüstringen Building Office, Garden Architect *Herrmann Foeth*–Cologne (p. 16 [66]), *Erwin Barth,* City Garden Director, Charlottenburg (Appendix 5 and p. 29 [77]), Garden Director *Ludwig Lesser*–Berlin (p. 31 [79]), Dr. *Werner Hegemann*–Berlin (Appendices 4 and 7),[3] Prof. *Max Läuger*–Karlsruhe (pp. 48 [95] and 51 [97]), the Garden Architect *Gustav Amman,* in the firm Froebel's Erben–Zürich (Plates 18 and 19) and *Wilhelm Schubert,* in the firm H. C. Haas–Wiesbaden (p. 39 [86]). The two designs on p. 15 [65] and p. 116 [156] come from my students and coworkers *Wilhelm Luserke* and *Joseph Kumpan,* the illustration for Plate 3 is from *Emanuel v. Seidl*–München. Appendix 6 was provided to me by the City of *Essen* (Deputy *Schmidt*), and Appendices 1 and 2 by the City of *Schöneberg* (City Building Inspector *Paul Wolf*). I owe special thanks to the worthy Central Office of Public Welfare [*Zentralstelle für Volkswohlfahrt*] Berlin for images and statistical data.

All remaining illustrations and drawings are from projects following my designs for the firm of *Jakob Ochs,* Garden Design and Construction–Hamburg, realized over the past ten years.

plate 1

Path with hedges of arborvitae; bridge with arches in the background.

THE PRESENCE OF THE GARDEN IN HISTORY

I

When the relationship of gardens to people is discussed in garden-historical terms within the numerous textbooks on fine garden art, this is always in reference to appearance only, to the *form* of gardens. By contrast, we seldom hear about the factors that were essential in *creating* and maintaining these.

For we who want to design new gardens, though, the style of old is of little use when taken literally. We don't want to imitate because we know that won't lead to a whole. Much more than that, we want a new flowering of the garden to come about organically in our own time. Form is always the end. For us at the moment it is more important *how* the gardens of old came about and *who* used them.

Let us briefly consider the names that appear so often when we study garden history.

In the Schi-King, a poem tells us about an emperor of China, and how he wandered happily through his extensive gardens: "The powerful *Prince Wen-Wang* in the wooded hills of Lin-Yo"—gardens of legend.

Just as legendary are the "hanging gardens of *Semiramis,*" and also just as legendary and mythic are the gardens of the *Hesperides*.

As a grain of truth is almost always found within myths and legends, so we tread upon even more historical ground when we hear of the famous gardens of the Persians. *Darios* is more favorably known through his so-called Paradise, enormous, richly varied tree plantations in desert regions, than for his wars.

In old Moorish *Granada*, still today magical memories are woven in among the blooming ruins of the garden courts of the Alhambra, and Bernal Díaz del Castillo, the conquistador, reported excitedly on the splendid gardens of the ruler *Montezuma of Mexico*.

In old Rome, it was the dominating class of the *patricians* who created luxurious country houses bordering the cliffs of the Apennines. In a letter to his friend, the famous historian *Pliny*[4] reported in great detail on his own ingeniously furnished garden.

A thousand years later in the same country and often in the same locations, the Florentine, Venetian, and Roman Republics brought a garden culture to fruition, the remains of which still evoke the admiration of all travelers to Italy today. Families such as the *d'Este, Albani, Pamphilj, Medici,* and many others have earned timeless reputations, not least because of their gardens and palaces.

And in France in more recent times, the kings and members of court again possessed beautiful, extensive gardens. We would not be able to form a vivid picture of the *Sun King* and his circle of *Richelieu, Fouquet,* and *Liancourt* without the strikingly intensive garden life of that era, which largely reflected their intellect, joy, and power.

Then there were other times, when circumstances beyond the ruling class led to the development of important gardens. A tremendously active and deep garden life often sprang from the *religious* or purely *spiritual* milieu. Thus, the garden only changed masters. In particular, it was the monasteries and abbeys that since Nazareth exerted an enormously strong influence, both directly and indirectly, on the garden culture of the host population. From the Rossicon Cloister on Mount Athos up until the present day, religious communities and orders—I'll only mention the Carthusians and Cistercians—have possessed outstanding and well-known gardens, to say nothing of the gardens of the popes and cardinals, which often had worldwide fame. The internationality of these church organizations always led them to obtain new plants and flowers, along with ideas for their composition, from the lush gardens from the Orient. Thus, these gardens were often very richly appointed. For example, these typically contained a physic garden, which in every period contained all the recognized medicinal herbs.

This *physic garden* and the so-called utilitarian *herb garden* of the Middle Ages appear to be the real fathers of the medicinal and botanical gardens, as they have been maintained and organized up to the present time. Special, complexly laid out gardens for schools were later created, such as the well-known English college gardens, which deservedly can be labeled "gardens for intellectuals." Here garden art is the repository for religious and academic tendencies. This is *educational privilege.*

As outsiders, we may also mention the garden culture of farmers, together with their priests, who, after freeing themselves from bodily and material dependencies, often achieved truly significant and original garden works in all civilized countries, as a direct result of their independent life situation.[5] Thus, by way of example, we have the creative character of earlier essential garden periods coming together.

These examples suffice. From these we discern one fact, namely, that it was almost always the *leaders* of the people who in all periods possessed gardens of rhythmic significance. The history of beautiful and rich gardens readily shows us an impressive series of feudal names connected with them, but when we review *bourgeois* tendencies in more significant garden epochs, it is much more meager.

In ancient times, only among the *Hebrews*, as well as among the highly developed garden culture of the Syrians, did the beginnings of a heightened garden expression appear to have been present on a widespread basis. The Bible gives indications of this, but nothing is authenticated.

At the time of the *Fuggers*,[6] who themselves possessed widely celebrated gardens in Augsburg, it may have been that, in Germany under the stimulating influence of the Hanse, bourgeois garden life had a noteworthy charm and prestige. Hans Sachs,[7] the Nuremberg poet, had already sung about the garden of a rich citizen of his father city:

> Because therein were planted
> Some trees full of fruit,
> Breeds of
> Bitter oranges,[8] muscat.

The enthusiasm for gardening in freely governed Holland was probably at an even more popular level, already having been brought to a relatively high state in the Middle Ages.

In any case, with these few beginnings of a bourgeois garden culture, one may not deny that the precondition and incentive were a wealth far exceeding the life needs of the individual, as well as a strong expression of garden art (Le Nôtre) that occurred earlier or in parallel.—

Such were the bourgeois gardens. By comparison, history offers us nothing on gardens owned by *lower classes*, and aside from some beginnings in the Middle Ages, there appears to have really been no necessity to present any form of collective gardens.

Thus, it is clear: in all periods throughout the world the higher form of garden pleasure was reserved for the powerful, for those who held title over goods and life. The right to a garden was virtually hereditary: possession and descent first made this possible. The gardens of the past were the expression of an elevated material and spiritual way of life. *Garden culture was class privilege*, it was aristocratic. Everywhere we see that the term "garden," whatever that may have meant, was almost identical with independence and a higher means of living. Freedom and wealth of a spiritual and material nature were what brought the garden into being. Higher demands, leisure, and refinement resulted in its form, and above all: elevated garden living was always an accompanying development or direct result of the *economic expansion* of the people in question, which effected changes in social status. First, on the foundation of this all-embracing social change, communal groups of spiritual and material unity were formed, such as among Renaissance aristocracy, the Jesuits, or farmers in Lower Saxony—only then did noteworthy gardens appear: *Garden culture is essentially a product of society*.

But we also see something else: up to this point, *Germans* were never among those who held the leading position in the ever-important area of garden life. Should this happen now? It *can* happen, if we fully understand the one *lesson* that may be taken from this brief overview of world garden history, and learn to apply it among ourselves with energy and determination: *Garden living, like any other phenomenon in the struggle of peoples, has always been an objective reflection of the social-political and spiritual structure of its respective era. Accordingly, the coming garden culture of the twentieth century, if it is a possibility, must also be the true expression of our own living conditions.*

How do these come about, is now the question: What did other generations before us have and not have, and how are we similar? How does the life of our period characteristically differ from those earlier?

These are the brief answers sufficient for our purposes here: following a long line of development in world history, *technology* was given to us as a product of the new natural sciences. And this further gave us *transportation*, that wonderful invention that brings together the languages as well as the physical presence of the most distant races, as though they were neighbors from the beginning. At the same time, with these great and often admirable things also come their opposing sides, such as: mechanization, depersonalization, devaluation of labor, big business and syndicates, trade unions, wage and tenement blocks, in a word, *capitalism*. Such are the things that, in comparison to our forefathers, we have to our advantage.

But we will indeed become more like them, if we think of our new *sports life*, of *hygiene* and related developments, which will continue to take hold among broader groups as a necessary balance to the particular way of living forced upon us. The delicate beginnings of a new *social life*, which we occasionally observe, are founded upon ongoing fundamental shifts among the social classes within the body politic, pointing toward related developments in earlier historical periods.

In principle, we can only compare our time period with others when we consider their *period of economic flowering*. And then it is clear: the intensified exchange of intellectual and material goods with all its side effects and consequences—these we have more abundantly than any known civilized people before us.

METROPOLIS AND GARDEN

II

That which captures and outwardly reflects our new, our actual life, is in fact our entangled urban culture with the whole formidable complex of problems it brings with it.

I have always considered it a completely pointless exercise when people condemn the metropolis. Large cities have always existed, and the energy of their concentrations of human labor, the sharp intellect and flexibility of their thousands of varied enterprises often have been the foundation upon which human cultures have risen. In short, whether the metropolis is productive or destructive—it's there. And as a strong people we must face up to the metropolis with dignity: enjoy its good and fight what is harmful.

Germany is the country of big cities, and when it is said to be growing, that means its cities. Fifty cities with more than one hundred thousand residents and a total population of about fourteen million lie within one area, more densely than in any other large country on earth. But that also means: almost a quarter of the population in the German Reich are exposed to that which we know as the dark side of big-city life: dwellings with too little sun and light, unaffordable rents, expensive bread and meat, alcohol and long working days, poverty among mothers and children.

Need breaks iron. And our urban poor, those misunderstood, out of natural necessity in the end create—gardens. Thus gardens, gardens for one and for all, these are what will alleviate the corrupting influences within our seas of houses. The frightening statistic that cried out to us in the Düsseldorf Urban Planning Exhibition,[9] "twenty-seven thousand infants die of consumption in Germany annually," would be unthinkable if every city possessed enough *gardens*. The metropolis needs gardens by *necessity*, and I also believe that the energy and desire to create these comes from the pure *joy* of gardens. Create gardens!

One may say that when a dense enough population is concentrated in one area, this in a sense automatically results in gardens and has always done so. It is a fact that the city, in both positive and negative terms, gives birth to the will to

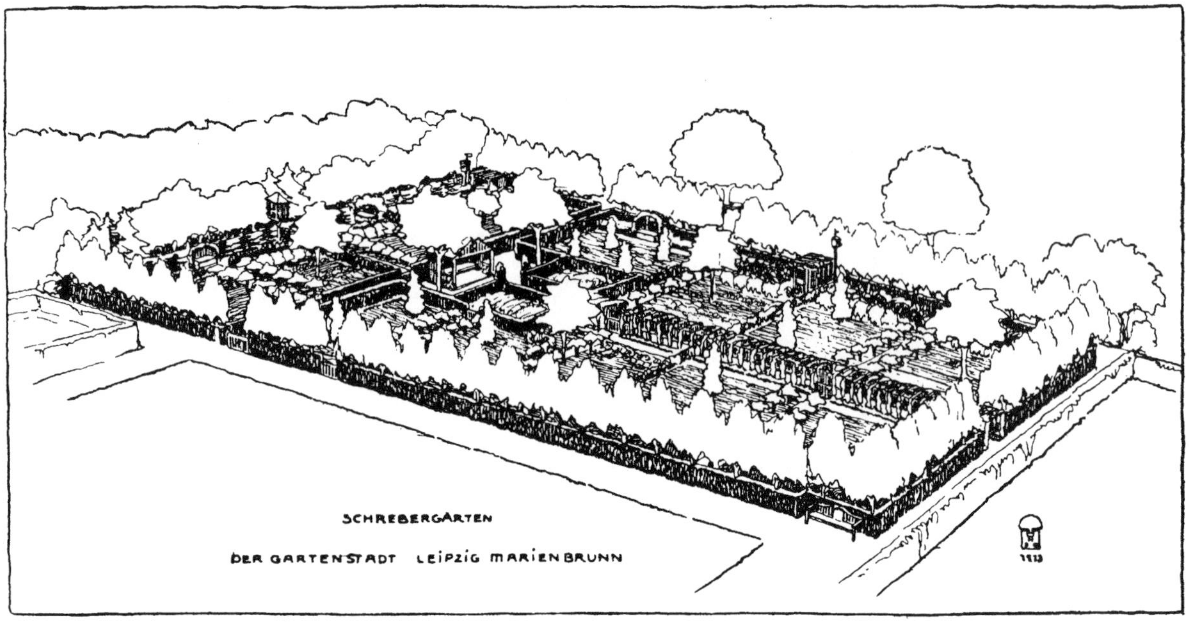

Schreber gardens of Marienbrunn Garden City, Leipzig
L. Migge

garden: this is also what has caused the creation of great garden epochs since the world began. *The metropolis a mother of gardens!*

When observing life among gardens in the metropolis, one may clearly distinguish between a number of effective arrangements. One group of gardens are those reserved for the use of *individuals* (private), followed by others that serve the *community*, and finally a series of *garden-like* forms found in the metropolis, such as the cemetery or zoological garden. In the end, all will be expanded and intensified within high culture, so to speak, through every essential effort, just as the *garden-city movement* values the garden above all else.

A. Private Gardens

ALLOTMENT GARDENS The metropolis as mother of gardens.—When our train approaches any large city from any direction, the first thing we notice is the blooming, often festively bedecked colonies of allotment gardens surrounding the noisy city like a green ring of peace. One thinks of the nerve-racking haste of the restless worker's body, one thinks of his still more deficient working conditions, of his various temptations, and then one will agree with me if I say: it should not be possible for any German city to be without this wreath of gardens; it is a *wreath of honor*: for every workshop ten and for every smokestack one hundred gardens!

Their origin and development can be divided into three categories of allotment gardens. The so-called *poor gardens*, the well-known *Schreber gardens* on community property, and the *workers' gardens* of individual factories and industries.

Among these the first one, in its many variations, belongs in the area of public welfare, as its name suggests. The people and the cities have recognized that among their poor, significant amounts of direct and indirect support only go toward alcoholism, tuberculosis, and other such epidemics, with their resulting physical and mental impoverishment. One therefore wants to seize this evil by the root, and preferably give the needy, especially large families, garden land for their own immediate use instead of their expected and indeed necessary financial support. One observes that it is much more important, and from a more enlightened viewpoint more desirable, to make downfallen members of the community into useful organs of the state and society through connection with the soil, rather than injecting a few hundred marks more from the public purse, taken from building land and taxes. Following this logic, one gives land for free or under cheap leases to the underprivileged. One wants to *break* poverty and gardens serve this end: this method of garden-welfare has been proven beyond expectations.

The *Schreber Garden Movement* actually developed from these beginnings, though today, as really more of a workers' movement, it may be called an all-encompassing social movement.[10] Once workers were able to recognize the various uses of garden living, they very soon moved toward self-sufficiency. The self-confident, well-paid highly skilled worker needs no poor garden! He thus constructs with the help of the local community so-called garden-pavilion[11] colonies, creating them through free self-help. Already in great numbers today, the prevailing allotment garden system appears to belong to the future. Our garden-pavilion colonies are spreading with a sheer elemental force. Thus *Leipzig*, the origin of the movement named after the local physician Dr. Schreber, can claim at present no less than fifteen thousand allotment gardens. That means nothing less than that in Leipzig sixty to one hundred thousand people of the lower classes may spend their free time in fresh air with the healthiest activities. *One hundred thousand people with their own gardens, that is a garden-social act of the first rank!* But in spite of this outstanding example, this progressive movement is uneven and relative as a whole, as proven by Berlin. Hemmed in by its suburbs, with only five hundred leased gardens, it only ranks among middle-sized German cities. (See Table 4.)

Forming an offshoot of these public allotment gardens are colonies built and supported by manufacturers and public administrations (railway and mining boards) for the betterment of their laborers and office employees. It is clear that through such contributions, these gardens generally make a more polished impression, due to their initial practical layout and long-term, mostly open-ended leases, than can the urban Schreber garden colonies, which with their sterile ground and hundreds of obstacles are often officially repossessed.

These belong much more to the category of the *bourgeois gardens* [*Bürgergärten*].[12] For the so-called higher levels of our workforce, such as office workers, tradesmen, and craftsmen who otherwise would not have their own gardens, here and there they have begun to establish leased gardens as a direct continuation of the gardens outside the city gates in former times. These are always of a somewhat larger, more solid, and more ornamental design in keeping with the greater economic power of this class. In addition, there also exist a considerable number of colonies and clubs with the goal of health reform.

The facilities of these allotment gardens are also new and functional. Land areas from two hundred to five hundred square meters at most (more for leased bourgeois gardens) usually maintain one or more playgrounds scattered about, which are fenced in for this purpose. Here the little ones tumble about protected from dangerous traffic, under the steady watch of their parents who are busily occupied in the garden. But the well-established organization of Schreber garden societies and others goes well beyond this. Very often on the metropolitan fringe and sometimes farther out, they have set up open and airily disposed estates with dairy, holiday, and swimming camps for weak children.[13] They hold parents' evenings and refreshing and informative hikes in the surrounding landscape. Indeed, these energetic country settlers even take advantage of the winter months, unsuited to gardening, through improvised ice rinks in the open and handicraft lessons in enclosed rooms. There is no worker more productive, multitalented, and more useful for the general good than the Schreber gardener.

Considering this situation, it must be alienating when many, particularly German city governments, still take an indifferent and even condescending position toward the allotment garden movement. Garden societies generally complain about the lack of concessions made by city administrations, in particular about overly expensive leases and above all about the short lease durations. And in fact if one sees—an often repeated picture—how the newly laid out, often artificially cleared or filled-in land areas with more or less (usually more) sterile soil are transformed in an unbelievably short time into a fertile oasis by the tireless hands of the garden colonists, then even as an expert one can pay full tribute to such achievements. And when one must helplessly stand by and look on as these garden pleasures are suddenly officially swept away by some schematic street layout or tenement block, destroying years of effort with one blow—then something in us may well up in angry protest against such an obvious disregard for productive labor. No, there *must* be opportunities for the socially valuable work of the rental gardeners, these true pioneers of the metropolis, to guarantee and thus to contribute to the maintenance and support of this extraordinarily happy desire to garden for the life our people!

plate 2

Two characteristic examples of allotment garden colonies. Above, one located farther out, on land with a long-term lease, which is productive, lush, and well outfitted; below, one that embodies the struggle against uncontrolled building, exploitative leases, and sterile soil.

PLATE 2

Front gardens
M. Wagner, Rüstringen

TRANSLATOR'S NOTE
Architect and planner Martin Wagner (1885–1957) was a close friend and associate of Migge. The two possibly met when Wagner was city architect for Rüstringen in 1911, or even earlier, when they both attended the German Garden City Association study tour in England in 1910. Wagner was later the city architect for Berlin, and was instrumental in helping Migge gain commissions in that city during the Weimar era. See *Martin Wagner 1885–1957: Wohnungsbau und Weltstadtplanung; Die Rationalisierung des Glücks* (Berlin: Akademie der Künste, 1985). The illustration here shows Wagner's design for a streetscape in Rüstringen.

The first condition for all these allotment gardens is that they must not lie too far away from users' dwellings. Their profitability and general advantages are irrelevant if they must be reached by hours of travel. Further, a lease for available lands is to be demanded for as many years as possible at economical rates. The allotment gardener can never be fully certain that he may begin to recover his labor and costs for the cultivation and maintenance of the plot if he doesn't already have in hand a lease for a minimum of eight to ten years. Thus the many grotesque images of garden houses and fencing, thus the common lack of outfitting of the area and a possible unkempt impression of the allotment colony as a whole, because the short lease duration didn't encourage just the opposite. On land held under long-term leases, the neglect of gardens is never to be feared. Even more as I have observed, the German worker is able to achieve wonderful things in an area that is generally still unknown to him: soil cultivation, an overall solid impression, and the functionality of small buildings, will be exemplary once it pays off in terms of *years*.

In the end, it seems necessary to break with the pernicious custom of transferring lands to an overall leaseholder, such as often occurs in Berlin and many other places. Through this typically bureaucratic measure, the gardener as subtenant must not only pay unusually high rents, but is also exploited by unscrupulous sellers of alcohol and supplies in the most shameless ways. Thus, it is obvious that the positive act of gardening is not only negated, but sometimes truly transformed into public degradation.

In order for all of these requirements to be fulfilled well and without harming well-intended parties, it is also necessary that in the future the *urban planner* must pay more attention to the allotment garden movement, foreseeing from the beginning the means and areas needed for the most generous utilization. Aspiring German cities with broad outlooks have every reason not only to passively tolerate but indeed to support with all their energy this elementary movement of the working classes toward the garden, which is such an effective counterbalance to the dark side of city life. In this way, we may do without the usual token welfare gestures. Not welfare payments, or not alone anyway, but *help toward self-management*. And when good deeds are generally spoken of, such as we have here, it's better to think of future returns than not. Who can say what deep, far-reaching effects this lasting contact with the soil will exercise on the modern worker, who has been removed from it for generations? Who can say what perspectives will be opened? This is a process of serious social and even national importance, with potentially bountiful results that can scarcely be overlooked today. Future generations will have the privilege of fully experiencing the significance of this multifaceted phenomenon, which we briefly describe today with the words: "allotment garden."

URBAN RESIDENTIAL GARDENS The residential garden in its various forms is really no more than an enlarged Schreber garden adapted to the human dwelling. Its most modest representative is the *front garden*. Today whole books have already been written about it. With the best intentions, but as I believe, somewhat misguided. Following our progress in garden reform, the rather overblown front-garden question will completely take care of itself. Fundamentally, it really does not belong to new garden design problems in the literal sense, but to urban planning, and in fact needs to be considered by the eye of the modern city planner.

Here and there, real old-fashioned front gardens still crop up, on cramped and often long-overbuilt areas in small cities and suburbs, requiring and supporting an individual treatment; but these only make sense when they can boast enough depth (ca. 8–20 meters). Here, naturally, their design is fully dependent on the personal goals of the owner. But new front-garden areas, particularly in the urban rental house districts of the big city, where they are most often located, are to be considered less from the interests of owners than those of passersby. The front garden of the *Landhaus* [country house][14] isn't of concern here.

Our new front garden is a component of the street image, when the garden is properly composed internally, and not just for display. Its purpose is to bring refreshing vegetation to the streets of our outer neighborhoods, and to help enhance and define the street image through forms and colors. The design of such front gardens is of course closely linked with the actual street planting, supplementing and replacing it as circumstances permit. In addition, the plan and facade of the house, which the garden lies in front of, is an influence. And finally street width as well, and above all sun exposure. One should in principle never place gardens in permanent shade, especially in areas affected by urban conditions. Since outdoor gardens in all respects already require sun, warmth, and light, so it is truly absurd to further burden plants in big city streets with soot, dust, and the stagnating effect of shadows. Here even grass and ivy require great expense in terms of construction and maintenance. Therefore, front gardens belong on the sunny side of east-to-west streets, or on a square. A true north–south direction usually allows vegetation on both sides.

Once these preconditions are secured, our front garden will be open to the street or enclosed with a low fence to ensure good visibility. If house entrances are elevated above street level, then the gardens are better embanked or terraced. Because rental houses on a city street are usually of the same or similar building type, it makes no sense to disturb the beneficial rest of this unity: *individual front gardens should be united to form green or colorful strips.* Thus on an east–west street, we plant two to four rows of Exeter elms or pyramidal English oaks, and globe Norway maple or locust, with strips of lawn, separated and bordered with low blooming hedges of spring spirea, Japanese quince, forsythia, and other shrubs. Ultimately in the outer, more open quarters of the city, we may allow ourselves a single-flower-color effect or distinguish between streets according to season, in which flowers selectively bloom. So here we may have bright red summer phlox, there lilac and white autumn asters bordered by deep gold chrysanthemums, with other streets almost filled with blue iris beds against bright white birches. Even with more limited means, summer annuals such as poppies, nasturtiums, and sunflowers may be sown. The possibilities that could be exhausted here are legion. And if one is so inclined, this blooming street prelude may be followed by coordinated color plantings on trellises and balconies, creating impressions that we can scarcely imagine in the context of today's sober city streets.

Aside from all that, the fact remains that front gardens should be considered while the street plan is being designed. An indifferent, inexpert treatment here can ruin everything from the beginning. Many hundreds of city streets bear sad witness to this. When carefully planned and treated as a unified street image, this type of garden is an advantageous enrichment and ornament to building block exteriors, as an essential contribution to the spirited approach to urban problems.

Another garden type that originates with the city is the garden of the private urban building, that is, *the garden court.* Sometimes associated with the front garden, in fact most of the private gardens of the inner city belong here: the gardens of hotels and restaurants, in short, any kind of green area reserved for personal use inside urban building blocks. There is a general belief that under these indeed unfavorable conditions, good and functional gardens may not be achieved. Thus as is always the case: even though not insignificant in number and area, these gardens still bear the stamp of neglect. But this need not be. For such locations, plant materials must be carefully limited to what will thrive there, and in general one must try

to complete the garden image with permanent materials, with appropriate plants, trelliswork, plasterwork, and garden ornaments. No situation is so limited in this respect that a garden concept cannot be realized: blooming beds in sunny spots, edged with brick paths, a vine-covered pergola, a tall tree or perhaps only a trellis on the wall with fences about—these are the main elements of a composition, their pleasant, practical advantages and generous overall effect often causing one to forget even the most beautiful gardens composed only of vegetation.

Garden court in a boarding house, Berlin
W. Luserke

TRANSLATOR'S NOTE
Wilhelm Luserke (1884–1970) worked for the Ochs firm in Hamburg from 1910 to 1911, then became the director of the firm's branch office in Berlin from 1911 to 1913. The building and courtyard illustrated here are thought to be on Kurfürstendamm in Berlin. See Gert Gröning and Joachim Wolschke-Buhlmann, *Grüne Biographien* (Hanover: Patzer, 1997), 236.

An entirely new flourishing of garden courts may come about through this path-breaking urban design concept for big-city living. There exist residential quarters with narrow, quiet streets. These house blocks have large garden and play courts reserved for the use of occupants. (See Appendix 1.) In addition, there exists so-called staggered building massing along the block perimeter, which encloses moderately high or even low-rise individual houses and enclosed gardens within. (See Appendix 2.)

In the heart of the city, high up in the air, we occasionally glimpse quiet hints of a garden that should occur more often: the *roof garden*. In Germany, it is has not been

Roof garden
H. Foeth, Cologne

TRANSLATOR'S NOTE
Ludwig Hermann Foeth (1880–1944) was a garden designer in Cologne and Berlin. See Gröning and Wolschke-Buhlmann, *Grüne Biographien*, 96.

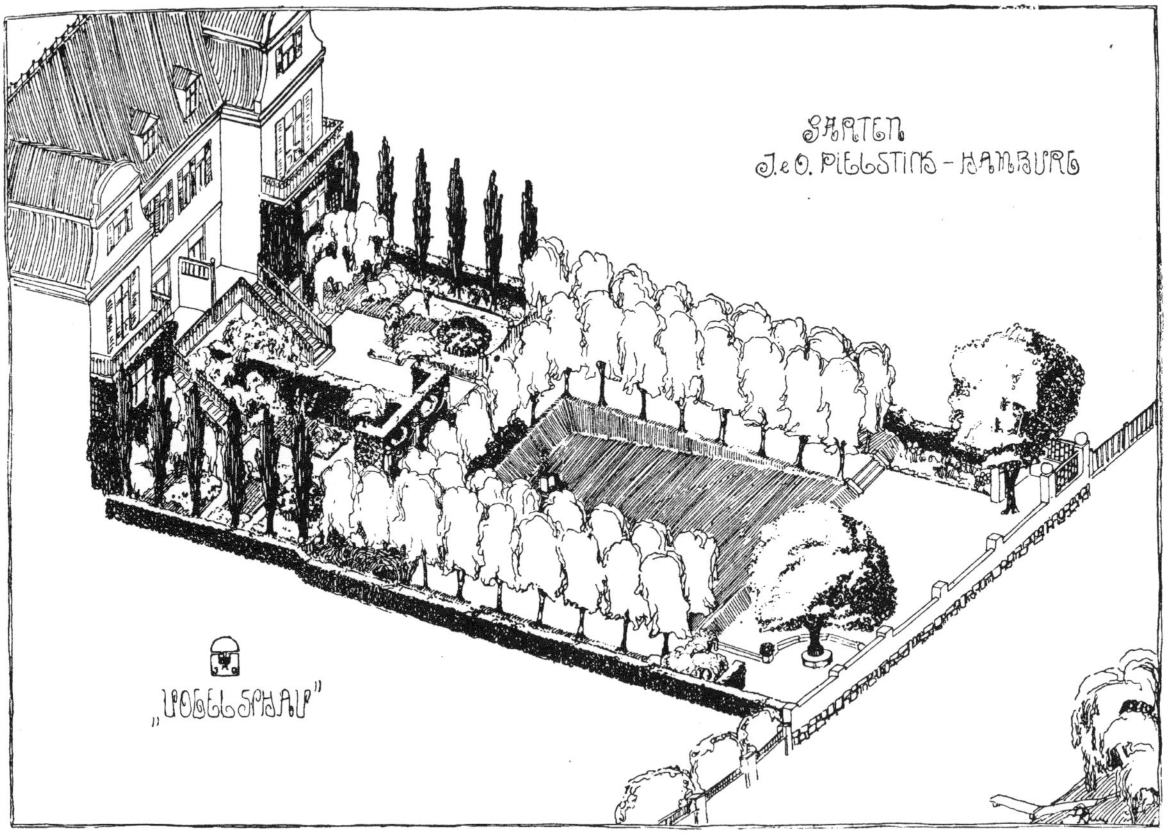

J. & O. Pielstick garden, Hamburg
L. Migge

Thiel *Landhaus* garden, Klein Flottbeck [Hamburg]
L. Migge

seriously considered in connection with urban problems, so that only a few incomplete examples exist. In America, by contrast, the roof garden is a widely accepted, highly used, beneficial facility. While we stick the mansard roof of the country house on our palatial office buildings at great effort and expense, and often to poor effect, the Americans go about the matter resolutely, crowning their skyscrapers in the truest sense of the word with areas for refreshment and recreation within richly outfitted roof gardens, for hard-working young people from offices and stores.

Yet farther out, in the residential areas of the well-to-do classes, we find a type of *back garden*, as a transitional form toward the actual *Landhaus* garden. One cannot say that these gardens, on a typically long and narrow, handkerchief-formed plot present an exciting prospect to the designer and owner. Wedged in closely side by side, and observed by the neighbors from every corner, one tries to make something out of it as is possible, and usually makes do. For these kinds of gardens, where permitted I have improved two gardens by joining them together. A partial coupling is also desirable here, where one may create a shared area for children's play and other functions out of the separate private garden plots. This latter method could be carried out more often, at least among sensible neighbors. As a result of land development policies, this type of garden cannot be completely done away with. Thus, one should just omit the ubiquitous but completely worthless and inconvenient private front garden, or connect it to the street.

LANDHAUS GARDENS A few more kilometers farther out we come to the region of *Landhaus gardens.* These are places of a private nature where modern garden life really may be pursued almost without limitation. Here is the El Dorado of the garden enthusiast, here he can indulge his passions practically unhindered. A lawn-tennis court, a rose garden, flower borders, benches and pergolas, as well as more unusual items, such as swimming pools or sunbathing areas, and even more may potentially be realized here. The general characteristic of the *Landhaus* garden is that it encircles the house with green. This means in fact two or three sides. However, I extensively describe these and other limiting conditions of this garden type and its particular design in the chapter on house and garden; here only garden living is covered.

The *Off-Site Garden.* This is what I call a garden type located off-site from the house, which is really nothing more than a transformation or expansion of the previously described allotment or bourgeois garden of the well-to-do. Our peculiar real estate and transportation situations today often drive even the wealthy toward this compromise garden. Anyone who wants to build a *Landhaus* in the best and therefore most expensive location cannot or does not want to make the extra outlay for a large garden around it. But in order to enjoy his highly cultivated garden interests more fully, he leases or buys a building on inexpensive ground, near his dwelling house—off-site. Here he may pursue his garden hobbies at his leisure and with pleasure, and with less trouble from the neighbors. Large vegetable gardens and orchards for household needs, rose, flower, and specimen gardens of the most

Reinhardt garden, Reinbeck [Hamburg]
L. Migge

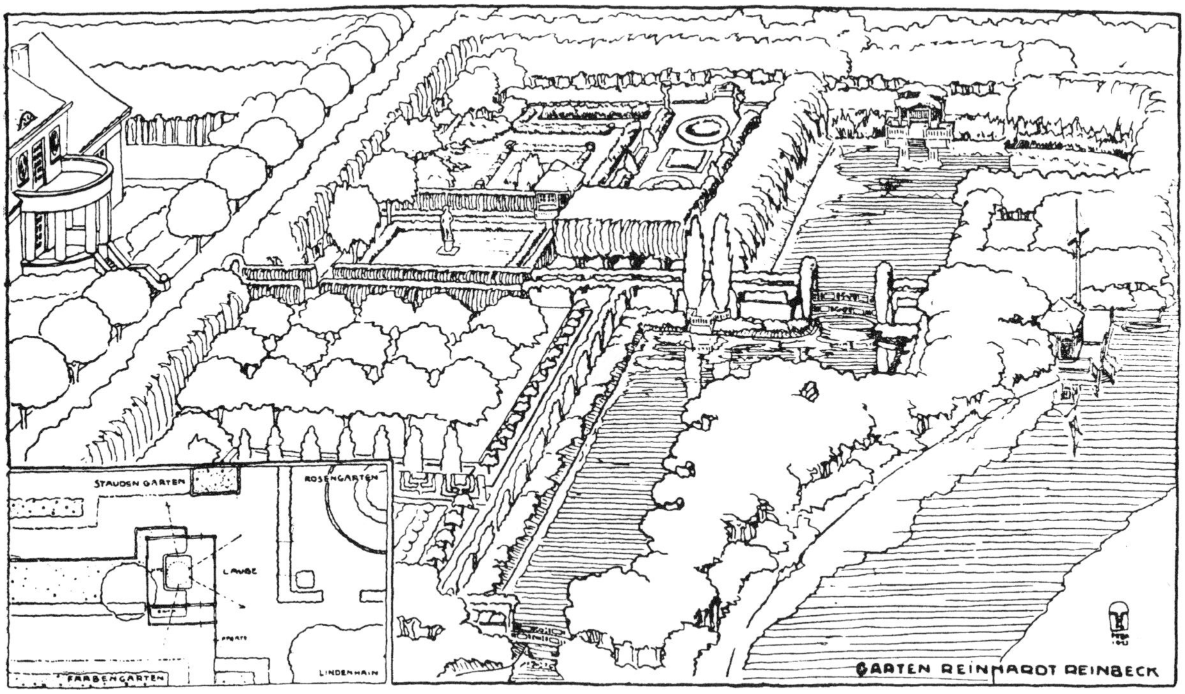

Park in Wohldorf [Hamburg]
L. Migge

varied kind find their proper place here. Here is room for experimenting with plants and improved breeding, as well as room for sunbathing and play areas. Because of the lack of economic pressure, everything here may be planted luxuriously from the beginning, perhaps surrounding a small house for shelter and equipment. I know people who through this tactic have created the most generous style of garden living, and who wouldn't have reached this standard if not thus removed. In addition, the generally provisional, all too provisional, garden-like arrangement on the more or less "building-ready sites" of our cities belongs in this category.

Finally as the last example of personal gardens we have the *private park*. To a certain degree, this is purely an external enlargement of the *Landhaus* garden. Since the park is more extensive, here I can repeat nearly without exception the garden facilities of the previous types, while also including more features here that are not possible in the others due to lack of space. Aside from the usual sport areas, football and baseball fields and, circumstances permitting, golf, hockey, and air sports fields may be laid out here. Even shooting stands and bowling lanes are not infrequent in this program. When the topography is favorable, there need be no lack of water areas for rowing and swimming, which may also easily include facilities for angling and raising fish. In addition, pavilions and shelters of various kinds may be envisioned for the comfort and pleasure of possible onlookers. For better viewing, terraces are provided with benches and resting places. Shady, wide allées, pleasant arbor walks, and sunny promenades must connect or enclose these good and useful things, as is to be assumed for a large garden. Following these coarser physical appetites that require the most space, the more refined sense of gardening is customarily reserved for the area closer to the house. Here diverse flower gardens and hobby gardening activities in all their variations have their proper places, made possible and supported by highly developed flower cultivation. Such special gardens may also be advantageously included within the other garden types, and will be discussed in the chapter "Vegetation." I don't wish to repeat myself. It's enough that one sees that for the owner of the modern park there are really no limitations on his garden passions. He can afford everything. But when this garden is well

PLATE 3

conceived, it doesn't only offer generous opportunities for sports and for refined socializing and flower cultivation. No, wealth obligates one in all respects. I cannot imagine a perfect park without a definite idea, without a rhythm, that would in a certain sense bring this multitude of things into one line, leading toward a monumentality. The highest level of ethics and taste, as has always been taken for granted by the privileged of a great people, must also be expressed in the contemporary park. The people should not envy the park owner, but must instead *observe* him.

UTILITARIAN GARDENS Within almost all of the previously illustrated private garden types, vegetable gardens and orchards have maintained a silent presence, and the smallest of these areas, such as allotment gardens, often represent nothing more than the utilitarian garden. Indeed, these well-known examples very instructively show not only the ancient processes by which a garden is made from an enclosed, practically laid out plot of land but also how it can be done *well* or *poorly*. Just as in the case of the hundreds of allotment gardens shooting up from the ground side by side, we see more dramatically here than anywhere else the entire range of garden values unfolding before us. Their expression reflects the whole person. Here, in these our most limited garden forms, one can make the most beautiful garden studies out of these utilitarian gardens.

But I only refer to this class of garden in this connection in order to counter the old prejudice that orchards and vegetable gardens are supposed to be rather ugly, indeed that they *must* be ugly because they have a plain layout. Among the Schreber gardens, those with the most practical arrangement are usually the most attractive. And we have enough examples of how these pleasant forms of orchards, vegetable, and kitchen gardens are easily elevated to the *ornamental*, when funds and inspiration are not held back (see Table 3).

B. Public Gardens

In the era of the general census, it is characteristic of the lack of a communal garden-will among us that serious statistics on public garden systems in German cities are still lacking today: garden policy is Hecuba to them. If we are to have gardens, then we are also to have garden statistics.

plate 3

Vegetable garden and orchard by Prof. E. v. Seidl in Murnau in Upper Bavaria, which shows how beautiful and creative such a garden can be, without ignoring the functional. The high, protective, useful walls make a significant contribution to very good effect.

TRANSLATOR'S NOTE

Emanuel von Seidl (1856–1919) was an architect who specialized in country houses. This image was possibly taken from his book *Mein Landhaus* (Darmstadt: Koch, 1910). See Joanna Waltraud Kunstmann, *Emanuel von Seidl (1856–1919): Die Villen und Landhäuser* (Munich: Scaneg, 1993).

It is otherwise in *England* and *America*. There they have truly public gardens, and proudly name them as such. The English people's parks, which we visit for study today, have developed to the fullest extent the principle of the practical use of green areas, and above all of nature. The English don't know the ornamental park in our sense of the term. Their practical nature hardly requires them to invent a particular form for their park needs. A section of the typical English landscape just outside their cities is merely enclosed, a basin constructed for bathing and rowing as needed, some paths are casually laid out in the area as required for access, a few shelters and a tea house are placed alongside these—and the people's park is complete. (See Appendix 3.) And even if not a shilling appears to be spent on luxury and show, there's still enough to please the eye. One sees namely the *people*, men and women, youths and children, who there by the thousands and more, lie about, walk, play, tumble, or debate on the lawns. The white and red of their garments plays against the dark green, just as brightly as colorful embroidery on a fabric of landscape. Their cries and celebrations fill the air. What need is there of ornament here? It is the people, *the living life*, that so truly ornaments these gardens! (See Plates 4 and 5.)

The practical, nearly tradition-free *Americans* are also far ahead of us in the utilization of public green spaces. We realize at what an incomparably greater scale the big cities of North America meet their park needs from the outset, when we hear that Greater Boston alone claims no less than 1,600 hectares of public land. And New York with its 340-hectare Central Park possesses the largest park area in the heart of a city anywhere in existence. In particular, the Americans have developed the small park type into a formal system. Their metropolises generally contain a great number of truly functional so-called play parks, instead of the few obligatory "ornamental areas" that are usual for us. Chicago, for example, in 1908 already possessed sixty-five such facilities alone, at four to twenty-five hectares in size, with club houses, sunbathing, showering, and swimming areas for both genders, running tracks, promenades, gymnastic areas, play lawns, wading pools, drinking fountains, and other lovely things, using up the last square meter. The Americans say: "The child without a playground is father to the man without a job."[15] (See Appendix 4.)

plate 4

The impressive, typical daily picture (above) from Hyde Park in London would be unthinkable in our public parks, aside from any other reasons, here because of the general lack of bathing suits; that below shows English children under the eye of their supervisors, digging in sand that was freely provided for this purpose.

TRANSLATOR'S NOTE

Carl Heicke took these photographs for a report on a study tour in England by German garden architects. See Richard Hoemann, "Erinnerungen an die Studienfahrt der Deutschen Gesellschaft für Gartenkunst nach England," *Die Gartenkunst* 10 (1909):167–68, 204.

PLATE 4

Those were our cousins. But how does it stand in our German Fatherland?

Greater Berlin,[i] this so richly privileged and blessed city, that we always have to look upon as the natural leader among German cities, in 1906 for its two million inhabitants counted three parks and 124 "ornamental areas," which together come to only 180.3 hectares. Even with the state gardens (the Tiergarten etc., 227.3 hectares), it possesses no more than *2.2 square meters* of green space per head. (See Tables 2 and 3.) *Hamburg,* the second metropolis of the Reich, can offer no more than *1.61 square meters* of public gardens. In contrast, *London,* that oft decried Babel-London, for 4.5 million inhabitants in 1902 already boasted 313 green spaces with 2,412.8 hectares. That means that for every inhabitant of this gigantic city there are 5.3 square meters of green space. And to the disadvantage of Germans this comparison goes down further, namely to one *square meter* (!!) and less per inhabitant, when we consider these three world cities according to areas of *accessible* green space. In order to compete with London (1902: five million marks), Berlin must make available about 3.5 million marks annually for the maintenance of its public gardens, that is to say, more than *five times* the current average garden budget. The worst is, however, that in this respect Berlin isn't alone among German cities when it comes to gardens. Hamburg's inglorious position is not much rosier, as already mentioned, even with the new city park now under construction. Bochum and Mainz likewise only show 2.5 and 2.3 square meters per head respectively, and two suburban towns adjoining Berlin, Wilmersdorf and Schöneberg, present a negative world record with their 0.9 and 0.5 square meters. (See Table 3.) Magdeburg, by contrast, is in a happier, higher position with 12 square meters per person, but such figures are misleading when they don't include *accessible* green spaces. By a similar comparison, large cities in England according to their category of density (and the German average city is not far removed), attain *3–16 square meters of usable public green spaces* per person, on average about 9 square meters of parks and play areas per inhabitant, and if the numerous private allotment gardens are included, then this figure is often multiplied. (See Table 2.)

These instances clearly show that here in Germany every practical facility must be left out of consideration when speaking of a responsible modern city park policy. Garden policy in Germany is as good as untouched. Our public gardens, such as we may indeed label as our collective property, were up to now essentially inwardly dead showpieces—"decorative green." In the best cases, these fulfilled hygienic purposes, but no more than that. Then come the obligatory gigantic areas for alcohol consumption, apparently inseparable from the larger urban parks. Indeed, that these often represent their most powerful attractions and their actual focal points, which instead should be about hygiene and morality, does not really make sense. There are people who by contrast recognize that this kind of cozy green architecture, this "rhythmic alcohol" in urban parks, this German *beer park,* is one of the most deceptive and pernicious enemies of the people. Even with such

i This and additional conclusions are taken from Issue 8 of the "Writings of the Central Bureau for the People's Welfare," Berlin, Carl Heymann's Publishers 1912; the tables included here come from this informative text, with additions by the author.

monumental architecture and cleanly shaven lawns with "unauthorized access forbidden by penalty," even with these ornate flower beds, romantic groves, and dank boscage, all of these being "commended to the care of the public,"[16] even if this pompous green stuff sometimes contains a modest playground hidden within it. Thus, today our public gardens still primarily serve a falsely conceived representation of cities. The out-of-town visitor long accustomed to their stereotypical appearance everywhere scarcely notices them anymore, and for the people as a whole their own gardens are truly foreign, despite the fact that these often carry the name "people's parks."

Our era, which has created a new coexistence for new people from the ground up—also contributes other principles to the new people's park. An aspiring people like us cannot afford the luxury of dead capital, whether economic or intellectual. Our masses don't want shrub and tree museums in the park that belongs to them; they rightly require facilities that are for active use, not just for looking at. The communal garden of our time will thus in the future be an instrument intended to satisfy the specific needs of the majority of people. The people should actively use them on weekdays and on their days off—we don't need *Sunday gardens*! People must be able to freely romp about in the new people's park, or it makes no sense. This is only a true people's park *because* its smooth and green fields invite people to *lie* on them, to *play* and *dance* on them, to *bathe* in their waters, and to *wade* along their banks. Only when the one-dimensionally trained and mechanically functioning urban worker is not only given the opportunity, but indeed is challenged after the day's burdens and toil, to turn away from work and in a sense be freed—only in this way does the people's park of today fulfill its true purpose.

Therefore, above all we must find in a good public garden those facilities that would be inhibited by the obvious limitations of a small private garden. First come all *possibilities for bodily movement,* then each positive opportunity serving the natural desire to learn among all aspiring classes, and further for the pleasurable observation of more flowers, sculpture, and good architecture. Curiosity and those minor stimulating passions induced by betting games, parade, and promenade are human; they can and should have their effect here. *Public gardens should be places of universal noble freedoms.* We understand today that play and sport, tumbling and gymnastics, indeed every physical discipline is indispensable for both genders and all age classes. Since all social classes suffer under high-pressure school training to more or less the same degree, then the requirement for physical freedom is also universal and prevailing. The question concerns the younger generation of our big cities. Yes, the well-being of our children, and thus the success and earnings of the future generation, I hold as directly dependent upon the *number of hours* that we require them to vegetate in narrow rooms, in corridors, or drafty, dusty streets and courts, when we could turn them toward free living in the fresh air among green and flowers, removed from heavy traffic.

Just how much our public areas lack the facilities for physical movement and any incentives for it, we may see from the simple fact that in 1906 among eleven Schreber garden clubs in Leipzig, the only representative of our social garden movement up to now, the enormous figure of 400,000 athletes could be counted on their playgrounds, yet the sport-friendly city of Hamburg in the same year could only muster 122,205. In addition, in Leipzig statistics comparing frequency of use clearly established that the Schreber gardens there, which cost the local government almost nothing, contributed much more to the people's health than the sum of the public green areas, and these required great maintenance. When we contemplate the full consequences of such figures and conclusions, then the problem of the shrinking military contingent in our big cities may no longer appear so inexplicable to us. And by the way, what happened at the Olympics in Stockholm in athletics that literally showed the popular power of the nations?[17] Following Sweden, why was it that in particular England and America distinguished themselves to their advantage (and to our embarrassment)? Whatever causes should be cited, I don't hesitate to state how it seems to me, that this was one of the most outstanding results of the uninhibited exercise of *life in the open*, which has been popular among those peoples for decades. *It was the one thousand play parks of the Americans that gave them their physical superiority and mental agility.* This is what qualified them, making them the young victors of the future. And you, my Fatherland? Oh yes, I know—you give birth to bureaucrats and professors for the world. Today, Germany officially runs a cult of the mind. But that's stealing from the capital of the people's power; it cannot and must not remain so. The day that the German people become aware that uniform sashes and standard poles, intellectual drill and fencing, barroom politics, hurrah and tarara-boom-de-ay alone don't make a people strong enough for the noble struggle to win the palm of culture in the twentieth century—from that day on, I say, begins the dawn of *our* park politics. Then we will embrace the thinking of our Anglo-Saxon cousins, in a sense *democratizing* the communal green with its whole range of significance for the people's well-being. Then the hundred thousand who until now have only experienced urban vice and absurd nervous stimulations will also learn of the true freedom that may be bestowed upon them. The hundred thousand will tread on this green plain[ii] every evening, just as naturally as they go to work every morning. The hundred thousand will learn to freely use their limbs, and to bathe their bodies in light! (See Plate 5.) There is no time to lose: we must make our public gardens *socially accessible* [*gesellschaftsfähig*] again, in the boldest sense of the term. On this point, we should emulate the English and Americans. We indeed will achieve even more when we organize this enormous field before us from the beginning, showing enough determination and courage to pursue a valuable model for positive improvement. Perhaps, who knows, it will again be left up to German thoroughness and largesse to contribute music institutes, public schools, and open-air

ii One has often heard that our climate forbids the constant usage of lawn and meadow areas by the public. This is naturally a question that belongs in the area of technology, which will promptly lead to a solution in our situation, as it has done under similar conditions. This question should not be cleverly exploited by professionals as a fundamental objection. In any case, in Cologne, Frankfurt, Hamburg, and other cities, we already have a significant number of green areas that are freely accessible for general use.

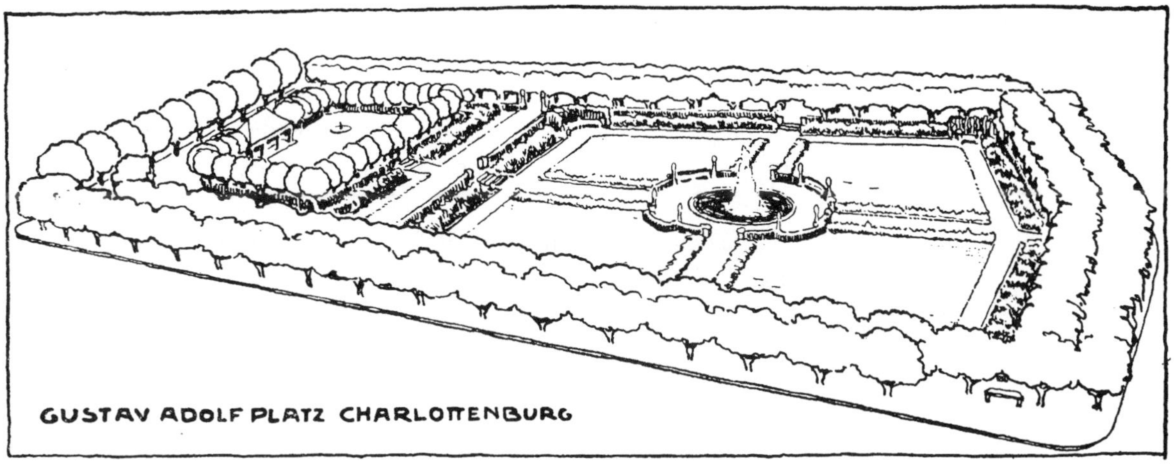

Gustav Adolf Square, Charlottenburg
E. Barth, Charlottenburg

TRANSLATOR'S NOTE
Erwin Barth (1880–1933) was garden director for Charlottenburg (Berlin) from 1912 to 1926. See Dietmar Land and Jürgen Wenzel, eds., *Heimat, Natur und Weltstadt: Leben und Werk des Gartenarchitekten Erwin Barth* (Amelang: Verlag Koehler, 2005). Gustav Adolf Square is now Mierendorffplatz in Berlin.

museums, as supplements to the sport halls and public libraries of the Americans, as possible intellectual additions to these green gymnasiums. This green plain must also be free for political and communal public gatherings; above all, it should be a factor in the public education of the German citizen. Yes, with an extensive enough area I can well imagine all of these facilities being surrounded by an open wreath of self-help allotment gardens—an improved type of bourgeois and Schreber gardens. The ideal people's park must return to every social class what the big city has taken away from them. *The socialization of the public green is one of the most important tasks of our time; it is no longer to be delayed.*

One-sidedness? This is not to be feared. It is of no harm when, along with these green functional buildings, the legitimate desire of contemplative people is fully provided for, through quiet promenades, protected resting places, and ornamental flowers. The necessary architecture can also be noble and exemplary, as we see with the new-styled developments in the American play parks. However, the fact that gardens and garden forms that are naturally derived from the daily life requirements of the future, and are marked with such a leveling social tendency, thus may also be impossibly banal in their appearance, is indeed self-evident. If the spirit of the largely formless contemporary English people's parks is to be labeled *naive*, and if we term the sober and unimaginative park architecture of the Americans simply *rational*, then may our later social park period be called upon to promise the highest values: it should be *monumental*. I hope that with us, the descendants of Goethe, that which is unique and new, embedded in such rich variety within modern park design, will be expressed in new, unusual rhythms, as though of itself.

THE URBAN SQUARE With its typically small area of about one thousand to five thousand square meters, this is an early, unmistakable product of the unparalleled density of our German metropolises. But in its current form it is hardly characteristic—except perhaps in a negative sense. The new methods of urban planning, still being defined, and the general change in attitude toward the purpose of public gardens mean that the location, size, and special design of this garden type cannot yet be more precisely defined. Here, everything lies in the future.

The great majority of our contemporary urban squares are at least breathing pores and green cells in the city body, but are ornamental in nature. These are meant to help *beautify* the city image through the most charming external appearance possible. They are in a sense the green towers and archways of the garden director, who wants to put himself in the limelight. However, I find that it would be better to furnish our urban green and ornamental squares with more facilities for practical use. They should not so much serve to screen off urban traffic as provide *refuge* from it. Refuge for pedestrians and especially refuge for inner-city children, who otherwise would be led to dangerous streets. Thus, our "lazy green squares" would consent to being redesigned as essential members of the urban community. They should be small children's paradises, with white sand and colorful flowers in abundance, as well as shade and shelter: urban squares should be *green traffic islands and flowery children's rooms.*

Only a fraction of these contemporary beauty spots of our cities, the so-called ornamental squares, should stay as they are. These, however, should be clearly distinguished in their outer form from playgrounds. And one should not hesitate and equivocate. Then these few, well-distributed symbols of the largesse and self-confidence of a city should have their blooming and decorative effects unified to monumental effect, rather than be scattered over a hundred places.

Urban square, Königsfeld-Kralove Pole
L. Migge

TRANSLATOR'S NOTE
Kralove Pole, today a suburb of the Czech city of Brno, was Brünn in Migge's time; extant.

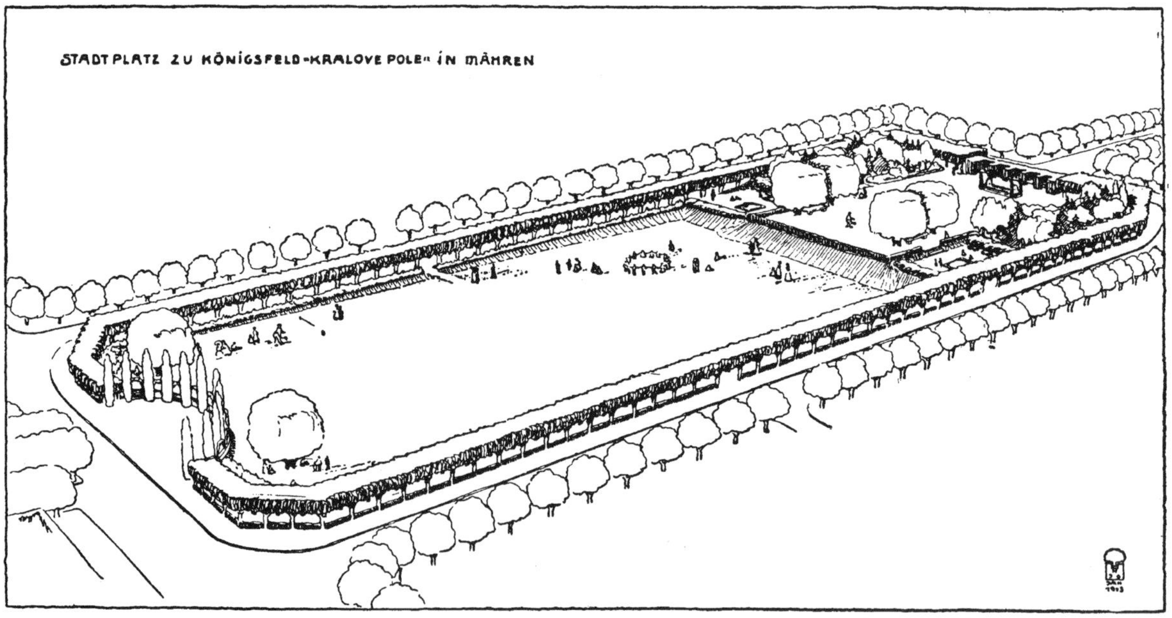

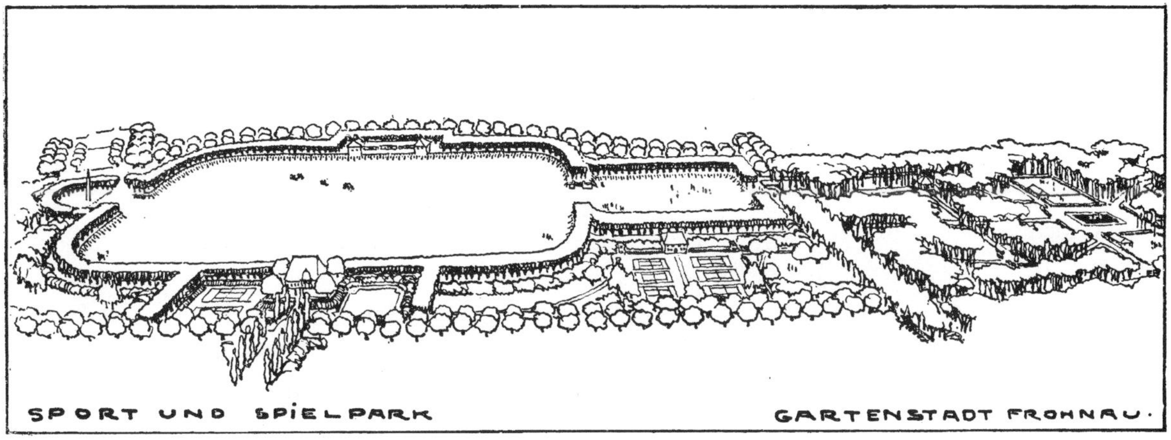

Sport park in Frohnau Garden City
Ludwig Lesser

TRANSLATOR'S NOTE
Ludwig Lesser (1869–1957) was a garden architect in Berlin. Frohnau Garden City is in Berlin; the park illustrated is now named after him. See Katrin Lesser-Sayrac, *Ludwig Lesser (1869–1957), erster freischaffender Gartenarchitekt in Berlin, und seine Werke im Bezirk Reinickendorf* (Berlin: Kulturbuch-Verlag, 1995).

GERMAN PLAY PARKS AND INNER PARKS The public play park, that practical American invention, I would really like to claim for us. In Germany we can scarcely boast of one of these. Only in Breslau, with its (1910) total of eighty-four sport- and playgrounds (ninety hectares dedicated to the public) do German play facilities show a somewhat recognizable effort in this direction. (See Table 1.) Following the best existing examples, one calculates for life in our cities one-and-a-half to two square meters for each member of the population, and for each child eight to ten square meters, excluding areas of school grounds, drill fields, etc., as occasional play areas only. The pathetic position of our capital city is shown by a study[iii] revealing that in 1910 only ten hectares of *dedicated* playgrounds existed—and these were only provisional. This directly results in the grotesque figure of *one-twentieth of a square meter per person*. In Berlin, where the kaiser is enthroned in the palace, the one whom we like to call the social kaiser, take note, *twenty* of its tenement house children together have *one* whole square meter of playground dedicated to them, and over two hundred thousand children there don't know even a speck of earth where they can really play. And in many other cities of the German Reich it isn't much better. Really, if this remains a permanent condition, then it's a disgrace!

If the urban square is usually not spacious enough, and when larger parks are farther out on the map, thus making the daily use of their facilities more difficult for everyone, then the play park of about ten thousand to fifty thousand square meters is needed to allow youths to pursue unlimited and varied opportunities for play. Therefore, these should be distributed in a greater number throughout the city (baby-carriage distance), especially in the outer working-class districts. And when ground must be purchased at high prices, there is still no reason to lose even one square meter of precious area to unessential uses, as this would be a serious offense to the taxpayer. Thus, the play park is simple and rational without reservation, open to serve the respective needs of neighborhoods. These needs typically

iii The reason that I cite Berlin so often is by no means because of "unavoidable dislike." To the contrary, I often admire the achievements of this city and its inhabitants. But in the end I make no apology for the fact that Berlin still today provides the best garden *statistics* and the worst garden *objects*.

require facilities for the amusement of the little ones (areas for baby carriages, sandboxes, places to run around, etc.), and for the older children and youth as well (gymnastic equipment, playing fields, wading pools and sunbathing areas, etc.). Occasionally there may be situations where functional reasons dictate the separation of these areas by age, and if possible even further by gender. There is no fixed scheme. This is also unnecessary, but the main point remains that as many *as possible* of these play parks should be available in each German city.

A more rigorous version of these play parks is beginning to crystallize here and there as the specialized *sports park*. (See Appendix 5.)

A different kind, distinguished only by its organization, is the *inner park*. It is a typical product of the most recent developments in our urban planning, and therefore still unusual here, whereas in England, as the so-called residential square, it has been generally common for a long time. The special feature of these public gardens is that their use is limited to neighboring residents, who occasionally contribute toward its costs. (See page 79, and Appendix 1.)

PROMENADES If urban squares and play parks are primarily for light-footed youth and the quiet play of the smallest children, then these urban promenades have almost the opposite requirements. Here there is no playing. Here one should be able to stroll with pleasure among flowers and under the shadows of trees. These are narrow, elongated gardens along a route, or are simply allées. However, rows of trees should only be planned where there is enough light and air for them to thrive. The still-common practice today of mechanically planting tall trees in narrow inner-city and commercial streets creates deformed allées that are a continual problem for city gardeners, and a sacrificial altar for the tax payer.

Most notably, such promenades exist today almost only as filler ornament for the ring roads of big cities. They are scarcely used, and only just manage to maintain their dead, although not inexpensive presence. These green thoroughfares serve their purpose much better when accompanied by so-called diagonal roads that connect the city center with the suburbs and the surrounding landscape. In particular, promenades provide the given necessary green connections between the green spaces of the city. The great expense of parks and squares is of little use in the end if they are only accessible through glaringly hot streets or overcrowded public transportation.

The width, length, and appointment of these pedestrian paths should be designed as appropriate to the significance of the places that they are leading to. According to their length, the more important ones must have paths not only for pedestrians, but also for bicyclists, as well as separate lanes for horses and automobiles, which under the right circumstances could be united to form a lively corso. When so designed, these promenades are also important members of the communal garden-organism.

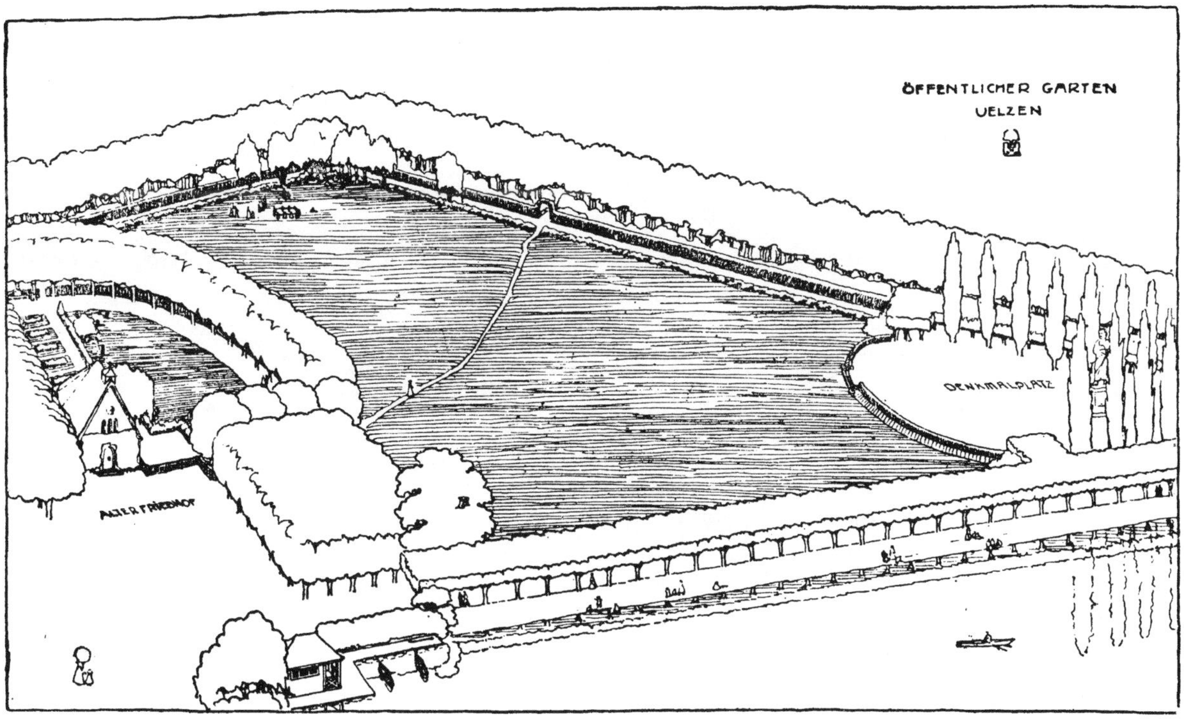

Uelzen Public Garden
L. Migge

OUTER PARKS These green thoroughfares gain more importance when a city conceives them together with extensive outer parks, as is of course to be desired. It is further desirable that these outer green areas, also called *city* or *people's parks,* are not too far removed from the city center. Thus, the worth of such park areas for the city-organism is higher the closer they lie to its heart. Since most dense European metropolises are *not* in the position to repeat New York's Central Park, and are not in the happy situation of Leipzig with its unique park policies, then the suburbs and outer districts will become the usual location for this type.

The outer park must also be reorganized. The traditional people's park type is not sufficient here. It must also be democratized, if it is to find a place within the great overall evolution of the garden. Here belong large fields for festivals and gatherings, there a jumping and running track, sledding hill, golf course, hockey and competitive sports fields, and over there large water areas with a beach (Wannsee![18]) and sailing and rowing lanes. One may also think here of nature preserves, concert halls, and exhibition parks. In addition, the need for varied and sunny paths, and for the observation of more beautiful, edifying vegetal scenes, must result in special pleasures, following by necessity from the definition of the term park itself.

This park form, with its occasionally massive number of visitors, naturally needs permanent buildings. As to whether the usual beer hall really fulfills this role,

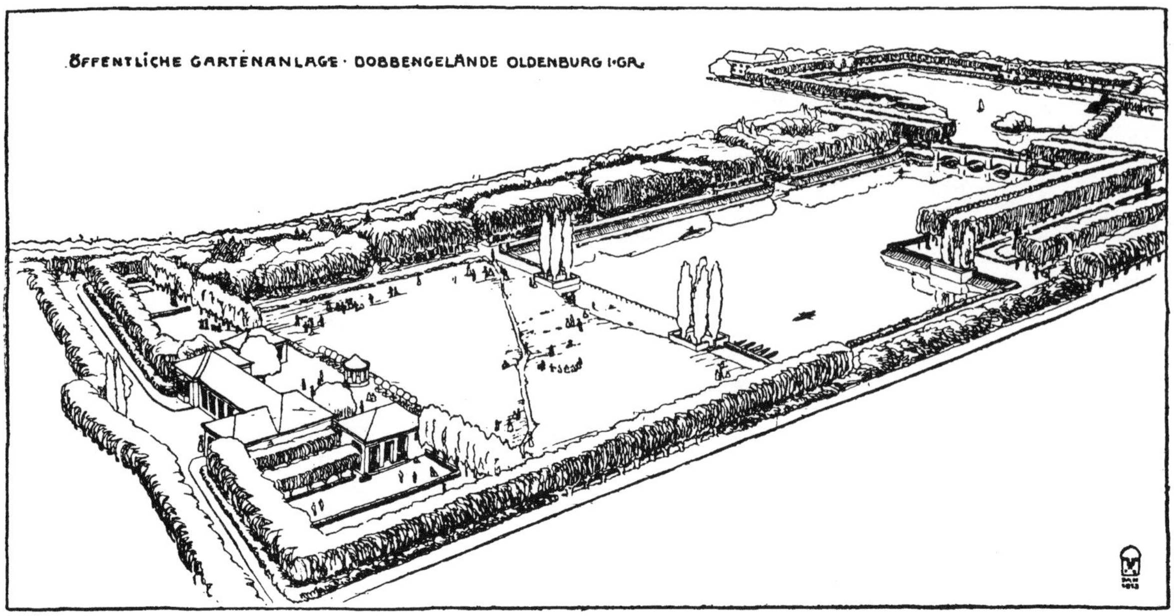

Dobbengelände
public garden facility
[Oldenburg]
L. Migge

I already have expressed my doubts. In England, as is well known, one finds simple teahouses with their chairs set out on the lawn under trees or colorful umbrellas in good weather. Only spacious shelters are sometimes necessary. It would be sensible for the gymnasiums, changing and club rooms, etc., which become more necessary with intensive use of park areas, to be combined with smaller coffee, seltzer, and milk bars. The interest and capabilities of our sport and social clubs could be used more effectively than before, by calling for their collaboration in the design of the park, hand in hand with officials and related organizations. Much energy that is eccentrically frittered away today (club fanaticism) would be retained for the community in this way. My park proposal for the City of Oldenburg is one of the first practical attempts in this respect—there, as an outer park, but for a metropolis perhaps nothing more than a play park.

OPEN SPACES This is a term that we in Germany still scarcely know. This refers to larger areas of farmland, fields, or woods belonging to the city, kept free from building on a long-term or permanent basis, or in which only limited settlement is allowed. These areas represent a permanent reserve from which the city claims its green spaces. At present these lands, purchased by far-seeing city fathers at reasonably low prices, are meant to secure freedom of movement in every respect for the city. And in the meantime they provide opportunities for welcome Sunday outings for the populace, especially for cities with uninteresting environs. Here and there spaces can also be left open for freely accessible park facilities and popular amusements such as open-air museums, ski runs, running tracks, festival grounds, and air sports fields. All of these naturally compatible elements that are now scattered about the city can be included in the area of the open space according to a preconceived plan. The generally randomly divided countryside surrounding our cities, alternatively composed of meadows and woods, always provides a good opportunity to harmoniously integrate such things.

The distant location of this great urban nature, on the other hand, means that garden forms serving day-to-day uses have no place here. This is why these parks exist. And for the same reason, all ornamental flower beds must also be excluded. And as with its use, the maintenance of these primarily agricultural and forested lands is also *extensive*. A few functional paths, roads, and other provisions for the control of traffic, partly necessary in any case, are fully sufficient to ensure the enjoyment of these lands for the public. Open spaces should not be an unaffordable burden on the city purse, but the opposite—its safety valve.

To my knowledge, up to now only *Vienna,* under its perceptive and active Lord Mayor Lueger, has truly broken new ground in the field of open space, through the creation of its famous wood- and meadow-belt. In Germany, cities such as Ulm, Leipzig, Gelsenkirchen, and others recently have proceeded to purchase open spaces, in modest but vigorous steps (city-planning administrations!). The city of *Essen* holds a special position in this regard (see Appendix 6) and appears to be developing into an increasingly outstanding "model city" for urban design. And most notably, it is the smaller of our aspiring cities, such as Rüstringen, Oldenburg, Flensburg, and others, that are showing some understanding for the important questions of park acquisition.

PARK SYSTEMS AND DISTRICT MANAGEMENT BOARDS While the previous kinds of urban gardens are the most important for general use, there is yet another series of related developments and transitional forms, which give rise to their own principles. But strictly speaking, a city with a good garden policy doesn't need to possess much more than a considerable sum of large and small play gardens and as many promenades and open spaces as possible in order to stand in a strong position in this respect.

It is not only good and practical that all of these garden forms are available for the metropolis, it is also necessary that they be intelligently distributed throughout the urban fabric following a carefully determined, all-encompassing *general plan*. Then only in the correct location will an urban green space obtain its optimal effect, regardless of what kind it is, and only the most direct connections to other similar spaces guarantee the most intensive use. In other words: open spaces, parks, promenades, and squares in significant numbers soon lead to a park system, for technical-management reasons alone.

In Germany, the comprehensive planning of green spaces is still in its infancy. While under some conditions our cities already have good communal gardens in

PLATE 5

sufficient number and size, they are usually isolated and widely scattered throughout the urban region; these came about *by chance*.

But the practical Americans go about it another way, as I must keep repeating. A stimulus for this may have come from the experiment in *Philadelphia*, where already in 1682 a system of urban squares was envisioned. *Washington* also very early on (eighteenth century) laid the foundation for a far-reaching organization of its park areas in connection with an overall city plan, at present being carried to completion. Columbia[19] followed. *Greater Boston* went even further, for under the direction of the private landscape architect Charles Eliot a system of open spaces and parks was created, with an area of four hundred square kilometers and a cost of twenty million dollars. (See Appendix 7.) One thinks: *eighty million marks* for the green spaces of one single city! Following Boston, an entire series of American cities have set up permanent park systems, and not a few have used a district management board for this purpose. But one alone outdoes them all: the new city plan for *Chicago* includes enormous garden areas and promenades (ca. 240 square kilometers), water parks, yacht basins, and powerful green urban axes, of a boldness never seen before, striking us in the face as a new, coming era of municipal power and civic greatness (Chicago—located in America).

These park and open-space systems of American cities contain everything conceivable for the relaxation and exercise of people in a dense metropolis. Much farther out, but still within the urban fringe, lie extensive parks, woods, and reservations, the future lungs of the expanding city, connected to one another and the city through miles of promenades and avenues. But in the inner city amid surging, noisy traffic, we find dozens of smaller and larger parks intended for shelter and ease of use, while ornamental gardens are only placed in the vicinity of public buildings and learning and art centers. Everything is directed by a grand, well-informed, and foresighted will.

These are creations that are far-reaching enough to bestow untold, blessed assurance to the future of millions of coming generations, creations that free them from the curse of living their lives among the stones of the city. Yet again—these are the deeds of youthful America. These powerful park systems were made possible through the economic and spatial planning of *park district management organizations*, these new, effective instruments of monumental civic thinking. The park district management board of Boston, for example, covers an area of thirty-nine

plate 5

Ten thousand tumble on the meadow of this American park, the whole of them undisturbed. The adults are in a typical "field house," in the library, the dressing rooms, and in the community, lecture, and entertainment halls. In front is the swimming pool, with tennis courts, sunbathing, and play areas to the sides and rear.

TRANSLATOR'S NOTE

The upper image is Franklin Park in Boston; the lower image, the boathouse/refectory (demolished) in Garfield Park in Chicago (source: Julia Bachrach, Chicago Park District, email to the translator, February 2013).

communities with a total population of 1.6 million people. In less than twenty years, the total sum of thirty-four million dollars was donated by generous citizens: working-class, middle-class, and wealthy people, all the same. One hundred thirty-six million marks for the green spaces of a single city!—may we ever hope to be able to match this yet unthinkable sum through public spirit? In the face of such achievements, the comparatively little that our cities have to show is truly depressing, regardless of how impressive they may otherwise be. We could remain in peace, if these were only *figures*, but they are infinitely more: these mean simply *strength. Within these figures lies a world of unimaginable cultural possibilities.* And one asks oneself involuntarily: How do we respond to this obvious lack of vision and energy? Do we still have any opportunity to have our say in the future?

O Barbarossa,[20] let a German Reich of proud citizens come, capable and willing, their lives worthily directed toward conscious goals!

C. Garden-Like Forms

A **CONCERT GARDEN**, for example, is to some degree a public garden, and yet not. For plants are neither the reason for it nor the dominant theme. A concert garden (restaurant, beer, café garden, or however its variants are called) exists for the musical offerings held there, or for the refreshment concession that encircles it.

The layout of the contemporary concert garden today still begins with the handling of the first requirement. Such a garden, for the most part dedicated to communal gathering, should above all else be laid out upon *terraces*. Each one of its guests should be able as far as possible to view (and observe) the greatest

Festival hall garden in Bingen
W. Schubert, Wiesbaden

TRANSLATOR'S NOTE
Walter Schubert (no dates); little is known about him. See Gröning and Wolschke-Buhlmann, *Grüne Biographien*, 351.

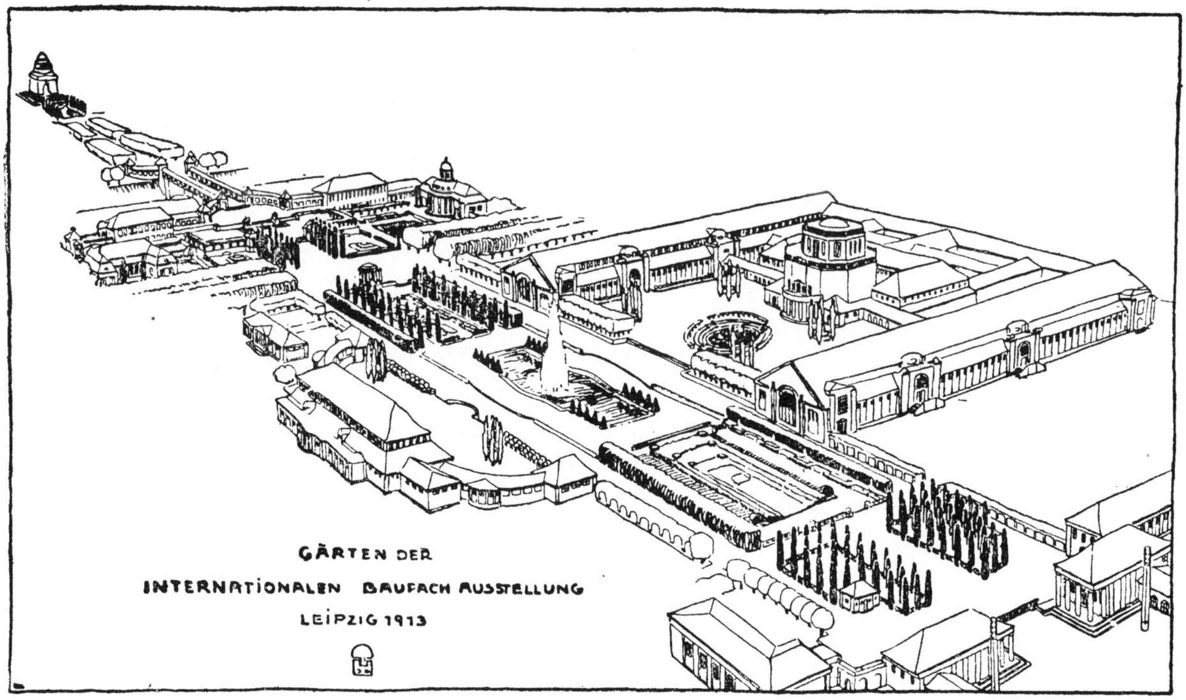

Gardens of the International Building Exhibition [Leipzig]
L. Migge

number of others there, yet in such a practical manner that even children can listen carefully to the sound of the concert throughout. Because we humans are all approximately just as tall, or similarly just as short, twenty to thirty centimeters is enough of a height difference. If trees are purposely planted to provide shade and shelter, the music pavilion acoustically well designed and situated, a promenade established nearby, and some ornamental flowers are provided, then such gardens may be desirable celebration areas, even if relatively small.

The facilities commonly created to provide for our exhibitions are usually not much more than expanded tavern gardens.

EXHIBITION GARDENS These may be just as varied as exhibitions themselves. The only thing they have in common, because they are exhibition gardens, is the temporal nature of their elements. In opposition to other gardens, it is conceived for a year or less. That means for the plants only *one* spring, *one* growth period. Here we have the particular weakness—garden bodies also have their illnesses—characteristic of this garden-like type.

For organizational purposes, it is generally not possible to utilize two or more growth periods for exhibition garden plantings. One must recognize this disadvantage from the beginning, and impose limitations on the selection of plants and the overall garden composition. In addition to old shade trees, walls, and vine-covered trellises, in the exhibition garden, which is always supposed to quickly "show me something," there should be many conifers and firs, periwinkles, flowering bulbs, and other plants with root balls, but only a few freshly planted tree masses and shrubs, which bud too late and too unpredictably. In unfavorable locations, tubs and containers filled with flowers, laurels, Buxus, or Taxus help to create a garden-like impression. Large quantities of annual summer flowers and vines, which according to my experience do better when cultivated in advance in comparison to hardier shrubs and other long-lived plants, contribute much to this. In addition, one must fertilize sensibly, and above all the soil has to be very carefully prepared. In short, there is only one consideration here: How and with what materials do I create as quickly as possible a garden that is temporary but that has the most dramatic result? The effects may easily be cruder than usual; that's the case generally with exhibitions, known to be relative successes.

Their organization must be allowed to follow the oncoming changes foretold by the celebrated garden designs of recent exhibitions (Düsseldorf, Mannheim). This has been valid, and not just for garden design exhibitions; in other words, the opportunity to aid the acceptance of new garden thinking through the exhibition of enclosed gardens and garden components was welcome until now; but there is no pressing need for this anymore. We have witnessed this, and have *fundamentally* decided. Everything else is neither the purpose nor the potential of exhibitions. The necessary intimacy of a house garden, for example, contradicts every attempt at its temporary representation, even on a technical level, when the openness of its setting has not rendered its representation impossible here in any case. Thus, the garden rhythms of an exhibition are organized to follow the special character of its architecture, economics, and transportation technology. Gardens become a *representative* factor within the overall undertaking, which will become less of an option the more we tire of exhibitions. Thus, its transformed appearance in unison with the previously discussed technical issues, and in connection with the new, generous tendency toward the specialization of our exhibition design in general, will result in a new garden category—the exhibition garden type.

For purely garden design exhibitions these principles are even more critical. Here *multi-year preparation* and the readiness of the area are essential preconditions. But even when these are secured, garden design exhibitions will only come into full blossom when they *specialize*, following the trends of the time. It is no longer sufficient here just to promise beautiful flowers and an abundance of green; rather one must seize a particular aspect of garden living and elevate it to a program. Only when they are supported by an irrefutable need and are concerned with current questions that are in the air will garden design exhibitions have a reason for being and be a great success.

BOTANICAL AND SCHOOL GARDENS Today these well-known gardens are already present in medium-sized cities as permanent facilities. They are not really

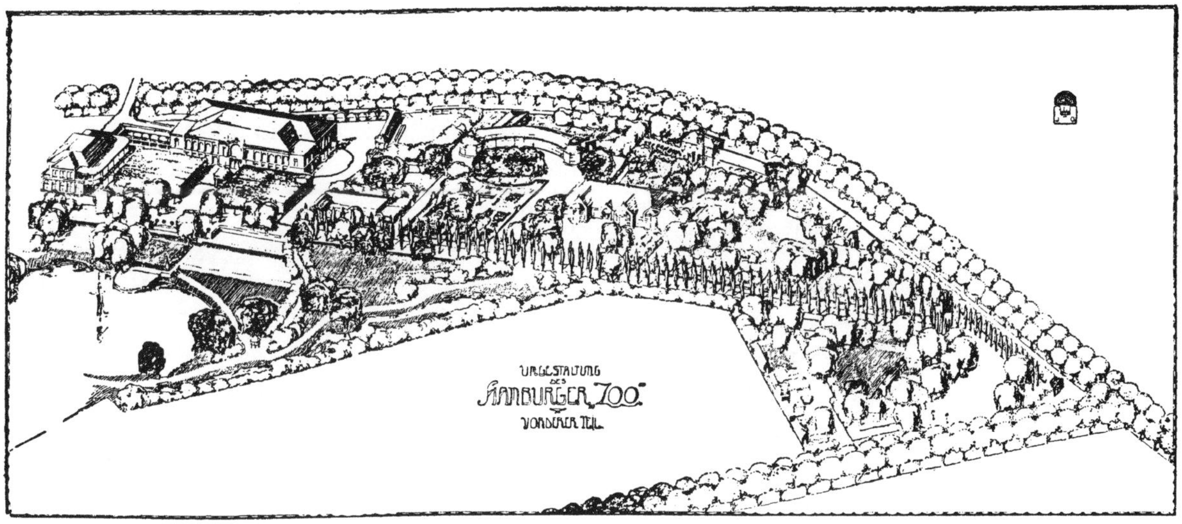

Hamburg
Zoological Garden
L. Migge

garden-like, but are really pure utilitarian gardens—pedagogical utilitarian gardens. Therefore, these facilities don't have much to do with any kind of rhythmic intention. Here it's all about function. And that is as it should be. Thus, everything brought into these gardens that used to be labeled "art," with the exception of organizational principles, could not really be called beautiful. Just the opposite. A rock garden arrangement such as that in the Hamburg Botanical Garden has a fully pointless effect in this respect. In the face of this irregular pile of detritus, one has only one certain observation, that this must have all been very expensive and that one has paid for it out of one's own pocket. The pedagogical intention is fully ruined in this kind of educational garden.

We should be determined that everything we surround ourselves with display its character as purely as possible. Then these gardens will almost automatically contribute to the harmony of our lives. A school garden has only these signs: *plant labels* and *beds* (see Appendix 8).

It is only one step from educational gardens to *gardens for school children*. Here the focus is on the fundamental educational problem of making the physical and mental benefits of the soil once again accessible to boys and girls in the big city. The children prepare their own planting beds with their own hands. The richness of nature's creative acts dramatically unfolds before them. They produce instead of just copying as they are usually only allowed to do. They become beings—let us see to it that every German child have his own bit of nature again! (See Plate 6.)

ZOOLOGICAL GARDENS The zoological garden primarily has pedagogical tasks to fulfill as well. It should *display* foreign and unusual native animals for learning purposes, to good advantage. That is its main purpose. In addition, it should provide facilities that allow for all the animals to be professionally cared for and scientifically observed. These conditions give birth to its form. The main question is simple: How do I create living conditions for *this* animal, so that it will thrive as well as possible for the longest time in our climate? And how do I best display *this* animal for the benefit of the observer? And finally, since we have nothing to be wasted: How do I fulfill each requirement most *economically*?

These necessities become most evident when we observe the animal enclosures that dominate the spatial impression of our zoos. In the first instance, these should be functional. They must therefore derive their formal expression from such needs. That is particularly the case for the materials used to construct the enclosure. Their colors and the type of enclosure should either contrast to or harmoniously complement the silhouette and color of the animal to guarantee its best display (relief effect). The resulting simplicity of construction and freedom from inorganic decoration would make these zoo enclosures inexpensive.

Plantings would have to follow exactly the same principles. According to the animal species, different kinds of hedges of trimmed and untrimmed blooming shrubbery would give a friendly character to the borders. Inside the runs as well, shrubs and conifers protected by wire could be added here and there, enlivening them and guaranteeing natural protection, for it should be observed that in our climate numerous exotic animals like protection from wind and dust but only a few like our cool shade. Most of our zoological gardens today suffer from a surplus of old trees and at the same time from a lack of protection for the enclosures.

With regard to the overall development of the site and its operation, the grouping of related species and the heightening of visitor interest should be considered from the start to make it easier to deal with the often difficult orientation. This is best realized through a rational organization of the necessary paths, open spaces, and allées. Adding to their dubious reputation, contemporary zoos actually appear to be labyrinths that can only be negotiated by locals without risk.

There are still a hundred things to introduce for the good equipping of zoos. Our zoological gardens could be educational facilities of the richest kind and ensure the purest pleasures—if we approached their facilities with *natural* associations.

But that's not what we do. The absurdities that one is forced to digest today in a single average German zoo are not for those with weak nerves. One need not make much of the usual "scenery" here, which primarily consists of harmless, rehashed ideas from old-fashioned garden design; they've just been adopted. For example: a ten-meter-high boulder enclosed within a cage (which incidentally has eagles in it), or an "Assyrian ostrich house" (probably Assyrian because the native building

plate 6

The children cultivate their own beds with their own hands, and then play games after work is done.

PLATE 6

Gardens of the Junior Officers' School in Weilburg
L. Migge

methods of ostriches still haven't been sufficiently researched)—such things are typical of zoos. There are also gardens with towering "dolomites" (that ring hollow like the chest of the consumptive when thumped on), there are "lion dens with protective steps" (as though the beasts sleep more poorly when five meters away from human eyes rather than seven), one sees "Japanese magic gardens" (which unfortunately have weak magic), and other interesting made-up things. And all this is offered with an unshakable, serious, professional emphasis: one knows that there is probably nothing there that doesn't have its effect on a certain element of the public. Thus, all of this is completely mad.

There is not space on these pages to do more than skim over the absurdities of our current zoo practices, as stimulating as the amusing overabundant material is. One thing, however, is certain: our current zoological gardens have absolutely nothing to do with the term "garden" in the pure sense. The new and serious progressive gardening has neither the opportunity nor the incentive to do anything noteworthy here, so long as the fundamental principles for the natural organization of this garden form continue to be ignored. This judgment on the whole remains valid, despite the worthy effort of Prof. Seidl[21] in Munich, in which there were some successful elements. The cultural object "zoo" still awaits a solution.

SOLDIERS' AND RAILWAY GARDENS The garden as the equivalent of goose-stepping and knee-bending? If one will, then why not? In any case, it is a good indication of the inherent power of garden concepts that lately even the ascetically oriented military complex believes that gardens may serve them to advantage. Here we see that the usual facilities of barracks, junior officers' schools, and troop exercise areas that were previously dead, drafty courtyards are now interspersed with all kinds of trees and simple green ornamentation—without causing

Gardens of the Michels & Co. Silk Weaving Factory, Nowawes [Potsdam]
L. Migge

discipline or drill to suffer in any way. Through the adaptation of goals that appear rather dramatic here, garden forms that are often truly unique have already arisen: soldiers' gardens, which properly enrich the life of our nation in particular.

But it would also be desirable if our other large state organizations, such as the railway and the post, gave some more space within areas under their control to the garden question than previously. Much more can be done in this connection with areas near train stations, post offices, etc. In particular, a generous provision of gardens for office workers and laborers collectively as well as individually would be worthy of these otherwise outstandingly directed institutions.

FACTORY GARDENS How such garden obligations may be correctly and generously observed is shown by the modern factory garden. It is an important contribution to the theme of work and gardening. The workshop garden, the newest type in our continually improving garden system, encircles the typically sober buildings housing our sober labor with appropriate green structures. It advantageously serves the recreational needs of workers on their break, and when large enough even provides for pleasant periods of diversion for personnel on evenings and Sundays. Thus, in addition to lawn areas and some ornamental flowers, in such industrial gardens we discover plenty of shady and sheltering squares, drinking fountains, outdoor gyms, and occasionally an open field for collective celebrations depending on the layout and type of operation. That such a garden form would have a different appearance when next to a steel foundry rather than next to a silk factory is understandable. All in all, the factory garden appears to me to be the most appropriate means of alleviating the current tension between employer and employee, serving both parties and us all.

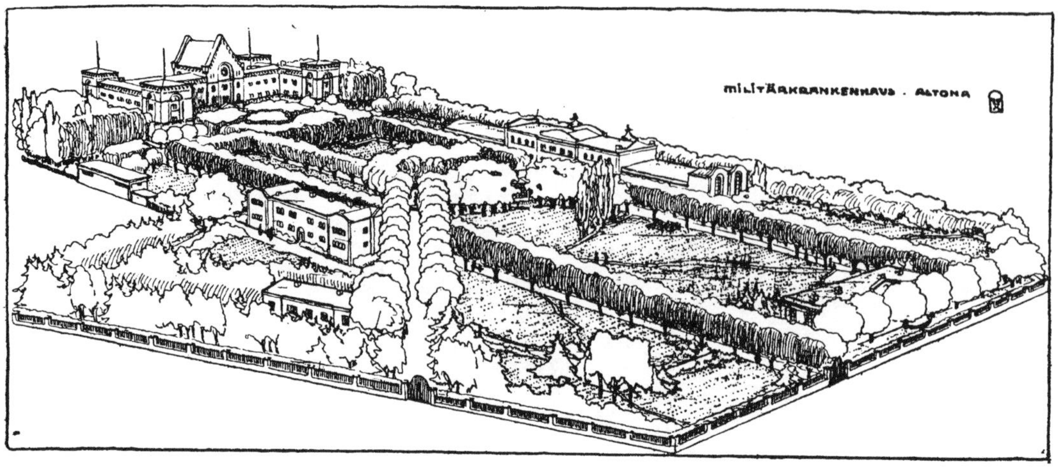

Gardens of the military hospital in Altona
L. Migge

HOSPITAL GARDENS This is also an important field of activity for the garden builder. The hospital garden as an independent organism is of relatively recent date. A place was first made for noteworthy garden design in connection with large modern hospitals with their spacious pavilion systems. And new principles for treating the sick, especially those in convalescence, mean that the garden is provided for hospital occupants as more than just pure decoration or external hygiene.

These days a complete hospital garden already responds to a variety of characteristic demands. Treatment of the sick requires many unusually wide paths for the movement of carts and stretchers. Large amounts of friendly decorative flowers must be used, and the patients' great need for rest must be accounted for with sufficiently pleasant sunny and shady seating areas. Protection from wind and dust must be carefully considered. Physicians and nurses require separate gardens and more. Even plant materials must be carefully selected for this location. A large number of evergreens are desirable in gardens for the sick, and hedges with nesting opportunities for feathered singers should be abundantly and appropriately laid out. By contrast, plants that are otherwise beautiful, such as laburnum, foxglove, etc., but that contain stimulating material in their leaves, flowers, or fruit are to be avoided. The sick are like children yet even more sensitive.

This garden category, to which belong sanatoria, educational establishments, and orphanages, in general has received too little attention. If life in the garden already gives so much to the healthy, then sun, green, and flowers may aid the recovery of suffering people.

CEMETERIES The question of how we should design our cemeteries is among the many apparent problems raised by our time, although in this case, a legitimate one. A problem that cannot be avoided by the most sober and most superficial. This time, it undeniably concerns *everybody*.

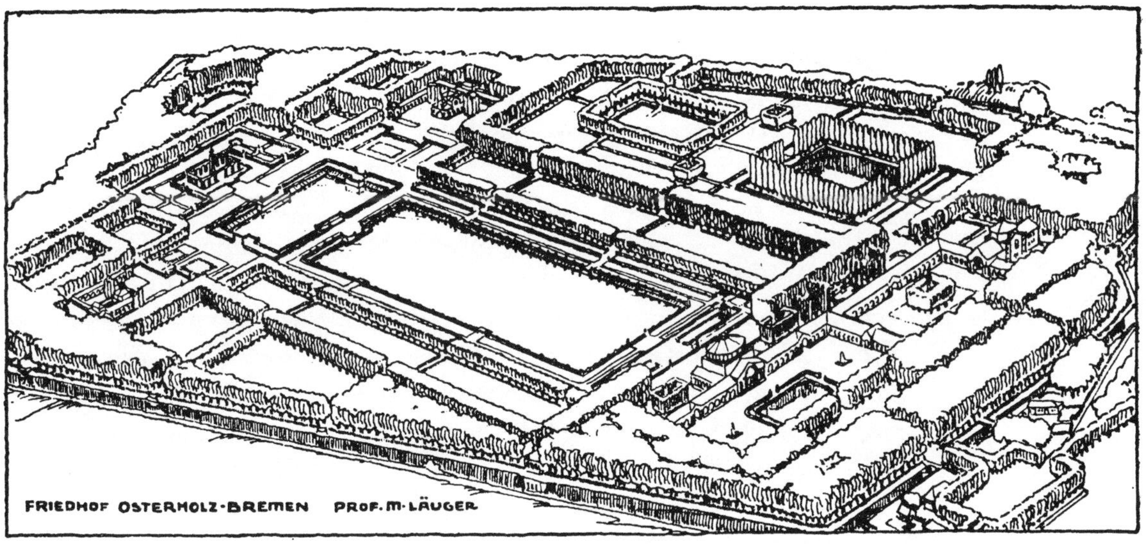

Cemetery project, Osterholz-Bremen
M. Läuger, Karlsruhe

TRANSLATOR'S NOTE
Max Läuger (also "Laeuger"; 1864–1952) was an important garden designer and ceramicist based in Karlsruhe; he was one of the leading protagonists in the garden reform movement and a colleague of Migge. See Rudolf Velhagen and Max Schumann, eds., *Max Laeugers Arkadien: Keramik, Garten, Bau, Kunst* (Baden: Museum Langmatt, 2007).

"It's all the same to me how I'm buried, just so it *happens* one way or another. Do I now have to struggle for my fellow human beings even during my pitiful rest, or, as a survivor, do I gain something for my body and soul by seeing my neighbor being honorably laid to eternal rest?" That's the usual talk. I definitely want to answer with an emphatic *yes* to these two questions. We millions of average people have little opportunity to have an ethical effect during our brief existence other than here, and as long as we still live nothing can scarcely move and free us inwardly more than to gaze upon the greatness and purity of death.

The power of the atmosphere of an old graveyard is undeniable. But this always depends upon the forms. This form originally resulted from the single act, the *cult*. In the end, it depends upon feelings and convictions as they are deeply rooted in us humans by religion and society. We are full of piety. A monumental expression can only be realized by means of the simple, defined clarity of its moral convictions. And: *as long as we are still not agreed among ourselves whether it is more conducive to eternal bliss to let the mortal remains burn in flames or slowly decay, then the great form of our cemeteries will likely be a long time coming.*

But even if we conceive of the contemporary cemetery purely as a burial site, some things could still be better.

The actual planning and care of cemeteries always has been handed over to the gardeners. However, the lack of clarity that reigns among professionals in this area is demonstrated by the paradoxically complex development of the cemetery in past years, which led to a transition toward a naturalistic landscape of various "mixed design modes." Only recently have we returned again to the original character of cemeteries, the *architectural*. We are on a track toward the correct, objective form for the place of the dead.

Following this logic, we quickly arrive at the main point, which alone promises a solution and also touches the spiritual heart of the matter. *This is the design of the single grave.* In spite of this, I don't believe that the improvement of monuments *alone* will serve this purpose. We must probe more deeply.

Certainly every healthy person strongly senses the disconsolate confusion of the burial areas in a big city cemetery. But when we attempt to clearly understand this perception, we realize that it is primarily the *variety of forms*, and the *juxtaposition* of the most varied monuments and grave designs that has this unpleasant effect. There is neither rest nor spatial expression. A field in a military cemetery with its sober, almost identical tablets and crosses, or even the rows of headstones of the Nuremberg Johannes Cemetery, have an overall effect that is satisfying, even inspiring. And to give an immediate counterexample: I find that our modern cemetery design exhibitions, as commendable as the conviction they evoke is, these mounds of good taste—their overall effect is not much better, nor more peaceful, nor greater than that of the fields noted above. If we translate this onto the mass-accommodations of our urban cemeteries, one cannot say that such means will really change their character fundamentally. It is not just a matter of *what*, but above all of *how* order will be arrived at.

Thus, we have reached the heart of the question of our cemetery design: *we need types.* Typical forms of gravestones and monuments. A thousand graves of the same kind. Here with wooden crosses, there with sandstone tablets, over there with iron memorials, always grouped in the same size and with an harmonious color scheme. Further on, a field with flagstones, an allée lined with sculpture, vaults, and chapels as terminal points. Urns in walls within arcades and halls.—Sober? Certainly, every great form is "sober."

And yet more. Gardening may also make a sensible contribution here. These fields will be enclosed as green rooms, using masses of trees and shrubbery at an appropriate scale for the human eye, here round and low, there by means of sharply defined pyramids. Each of these beautiful spaces is accented by its special dominant flower theme, its forms harmonizing with the grave types. A garden emerges from this. There we have the garden of the dead with red and white roses (or ten varieties, all different), one of white lilies against dark ivy, one of blue and yellow iris, and another with dark red poppies. And this one, truly colorful with blue forget-me-nots, violets, crocus, and periwinkles—this belongs to the dear little ones. And yes, citizen, you may be shocked I am doing this—I now chose a place of rest for myself in the garden of orange nasturtiums and sunflowers—brutally yellow sunflowers! The paths leading up to this feast for the eyes should complement it and prepare one for the experience. One thinks how lilacs, laburnum, and wild-rose walks, along with hawthorn, flowering apple, and locust promenades, alone could provide orientation. I find that one frequently becomes lost in most of our large gardens. And even more so in our gigantic cemeteries.—

Schematic section through the type—cemetery
M. Läuger, Karlsruhe

This is a small sketch of the image of a future cemetery. More detail is prohibited by the limited scope of this book. The cemetery of the types! When people are able to think this way, as they themselves long have been outwardly—*typical*, then the solution to this and many other questions would be significantly closer.

When, finally, will the poor, rich people of the capitalist period, we, the restless, harried creatures of iron and transport—when will we attain the true place of peace!

D. Garden Cities

We know that the situation of a city relative to its green spaces entirely depends upon its urban plan, affecting how well or how poorly the contemporary metropolis is able to enact measures for the future and direct its own development. And because the city must incorporate new areas from surrounding open country to meet its constant expansion, so the question as to whether the city may and can have numerous well-designed gardens is also a question of *land policy*.

The new garden-city movement has focused precisely on this point. It takes up the task of fighting against the land speculator's negative effect on housing provision by constructing garden settlements for the less well-off, on inexpensive, nonsalable land, with speculation permanently prohibited. The kind of urban forms that result from this socially commendable work, and whether these join directly to the city or strive to be fully new urban organisms, is scarcely our concern here. We are only interested in how and by what means garden cities embrace the problem of the garden, and how they have carried this out up to now.

Theoretically our contemporary garden cities are indeed well provided for. Here everyman truly has his own garden. But unfortunately this often remains in fact theoretical. More often than not I have seen that the gardens in many garden cities are poor, cared for without expert knowledge and without any particular passion. From the outset, there doesn't seem to have been much weight placed on solid technical facilities. The overall impression is that the leaders of this and related movements have not been able to devote enough *time* and attention to the garden question up to now.—

The term "garden city" for us today clearly stands for a building method combining perfect craftsmanship with a wide range of new domestic technologies. The characteristics of the garden-city dwelling are thus novelty and exemplariness. Only the term garden city itself reflects its origin, in the sense that it ultimately represents the full completion of living quarters through *gardens*. Following this logic, in fulfillment of the meaning of the garden-city movement, these should by no means be subordinated or treated as an aside; rather garden and house must be provided *at the same level of quality*. Accordingly, the gardens of a garden city can only be good if they are *technically perfect* from the beginning. Furthermore, the search for typical layouts and new garden forms should express the full sense of cultural responsibility among the overall community as embraced by the garden city in all of its activities.

By and large, the "garden-like character" of the garden city simply repeats in more concentrated form the previously described process of garden development in the city. The influence of the latter can be divided into two directions: urban planning and domestic technology.

The *planned organization* of a settlement in the country can and should be supported through horticultural means, in all the areas that are subject to communal use. That is, streets and public areas. Here experience has shown that all garden areas are best treated from unified viewpoints and from one position, in such a way that the character of a street, as determined by the location, orientation, and variety of house types, is sensibly complemented and enhanced through the type of plant growth. To that end, the typically cramped front gardens that result in an arbitrary, disordered treatment of the street composition should in principle be avoided and even fundamentally forbidden for north and west orientations. Their relatively high fences, necessary for private (not public) use, give the impression of plant cages that obscure the rustic or carefully carved stones of the base of the house. The space taken up by these thoroughly expensive front gardens would have been better served with narrow strips of lawn, rows of trees, or harmonious flower plantings. Thus one street could have displayed two rows of globe Norway maples, Exeter elms, or locusts on narrow strips of lawn, and another single row of pyramidal English oaks. Trees with large crowns are less often to be recommended, if left untrimmed. Such street images as these (similarly to those I have already described in the section on front gardens) thus assist and admirably reflect the garden value of a settlement incomparably better than front gardens themselves. The resulting unity of the individual streets at the same time differentiates them from others, contributing to overall orientation within the settlement in the most natural way imaginable. When there is not enough room for espaliered fruit, vines may be allowed to cover the houses to help reinforce these unifying effects. With what pleasant feelings might we walk down a Honeysuckle or Rose Street, or even a "Blue Clematis Street." For walls in shade and in unfavorable locations, ivy and wild grapevines serve well, and for especially protected areas common grapevines and wisteria are still good enough. Garden cities must be more "vine-covered and espaliered" from the beginning. One must not forget that our fast-growing vines are often the most preferred means of evoking a garden-like impression. And in terms of overall organization, these will always find a form, and cooperative planting among street residents is not excluded by this treatment, if in fact the community has not already decided on this.

Even more important than these communal questions is that of the *individual* garden. Both categories are mutually relevant to the grouping of houses. How should the house be related to the garden? Obviously the garden belongs to the sun; it must thus open out toward the south or east. Gardens can only grow this way by uniting the highest possible yield and utility with a beautiful appearance. This principle should be observed even when architectonic-spatial concepts and traditional urban forms appear to contradict it. The term "garden city" must hold so much duty and faith within it that one may simply wait while new forms spring up from these new conditions of use.

Even with the initial design of the houses, espaliered and creeping vines could really be taken more into account. In addition to the usual loggias and balconies, there must also be small terraces, verandas, or even latticework structures present everywhere in order to connect the house to the garden. *The small house is to be designed with respect to the garden from the beginning*; the architect and the gardener should work hand in hand from the first stage.

The *enclosure* of the gardens could be defined clearly by living hedges that offer better protection from dust and wind, more often than is done now. A small, broad-meshed wire fence about fifty centimeters above and twenty centimeters below the ground would unobtrusively protect gardens from rabbit and hare damage. If wooden enclosures are used, often the simplest fences of rough wood are enough, which may then be covered with abundant blackberries, wild rose varieties, and other conservable fruits. Only dwarf or espalier fruit trees should be planted in the small garden, and because of shadows only occasionally full- and half-standard fruit trees. In a garden city where the variety and newness of all the latest requirements demand overall economic planning, light and shade must also be organized. On the other hand, berry shrubs of all types, as well as hazels, medlars, quince, and finally ornamental black elderberry shrubs may be planted to advantage.

The solution to the problem of the *means of access* to individual gardens is equally important to both settlement planning and gardening. The concurrent use of these as service ways and secondary circulation is to be recommended; these should be unified in a formal system.

The question of *human material* is also not unimportant for a garden city. Poor pensioners and cultural reformers probably won't arrive at the solutions to these difficult problems on their own. Most of today's garden-city residents don't come from the country. The laying out and planting of gardens is new and unaccustomed to them; here they lack even the smallest measure of basic experience and understanding. Often, even the desire for garden living must first be awakened, as dramatic experiences in England have shown. For these reasons, the setting up of stimulating and informative *courses for horticulture* in garden cities seems necessary in order to achieve something other than the indifference that results from uninformed efforts and wasted energy. At the same time model gardens are to be planned, such as I have encountered in English industrial workers' colonies among other places. Once a basic understanding of horticulture has been attained among former city dwellers, then *just at that point* a virtual mania for originality in garden design tends to arise. In order to further support the ripening of this sure sense of gardening, the structure of the individual garden (main path, fruit trees, and enclosure) should be provided immediately following the construction of the house: the garden type is to be defined *together with* the house type.

These considerations automatically lead to suggestions regarding technical management: where at all possible, it is recommended that a practical *gardener* and fruit expert should be hired at the cost of the community, to permanently oversee and advise the small garden owner and to give simple demonstrations on technical matters as questions arise. If the size of the community doesn't allow for that, then a cooperative society of neighboring settlements could assist. This kind of cooperative could establish a small *nursery* on an appropriate, prepared area, with a connected early culture garden (bedding flowers, etc.). Here I can imagine all plants being trained and tested, as appropriate to the particular climatic and material needs of the settlement. Laying out the public areas and the structure of the small gardens and maintaining everything would be further tasks of this nursery. If there are no suitable private nurseries in the area, then local residents could obtain good plants here at reasonable prices. This practical garden center could also give advice on problems such as the practical, sufficient sourcing and application of fertilizer (composting, amount of artificial fertilizer coverage and its distribution, etc.).

Certainly there is yet a whole series of tips for the flourishing of gardening in a garden city. Thus, the introduction of prizes, small cooperative exhibitions, children's flower days, and much more. The financial incorporation of all these measures within the budget is also not of secondary importance. But such suggestions can only be made following a more exact understanding of local conditions. For now it is only necessary to understand that our garden cities must determine to take decisive measures and make fundamentally great sacrifices for their gardens if they want to ensure success. *This more* of personal and general effort should not be understood as an embarrassing over-budget luxury expense, but as the indispensable, self-evident precondition for living in open country—*that indeed is the future, complete garden city, as a city of beautiful gardens.*

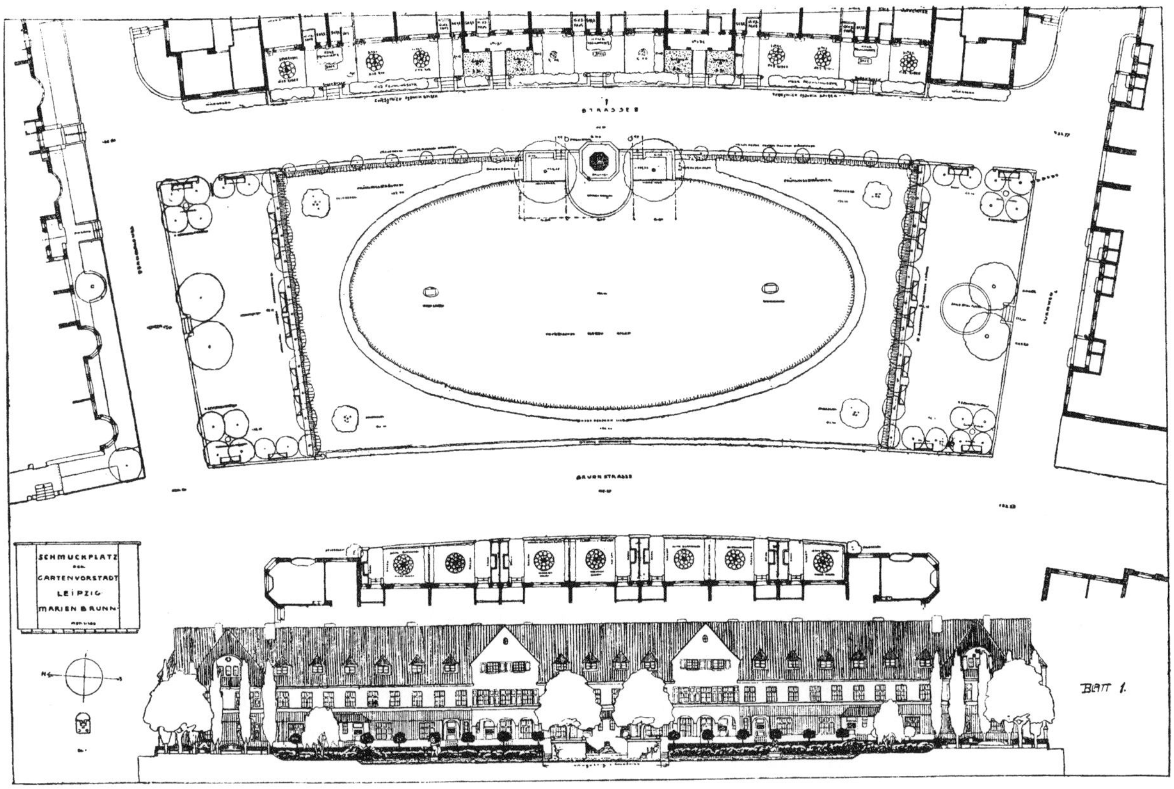

Ornamental square, Marienbrunn Garden City, Leipzig
L. Migge

Thus, we have an approximate summary of those basic elements that the metropolis contributes to garden living, whether desired or merely tolerated.[iv] People with mature minds and higher morals perceive the destructive side effects of metropolitan growth. They neutralize these destructive influences through their knowledge and means, thus bringing this oblivious human race that threatens to wither away between the stones back to ancient mother earth: *one makes gardens.* Societies make gardens, city and state and every individual makes gardens. Allotments and house gardens, playgrounds and parks, colonies and woods are created. Everyone has fresh air, and space, and rest, everyone has a piece of garden life. Energy and lust for living should burst forth; laughing and singing, freedom and pride, these should no longer be the privilege of the great!

Thus is born the great city of the future, permeated and bounded by green. It comes forth and presents itself as a happy balance between the skillful organization of work within and naive nature without. And the garden plays the mediator. Speak no more of big city poor and vices—flowers and green will heal them: gardens are the reaction of the metropolis.

iv I have already foretold this entire development of our gardening, which is beginning to take on practical form, in 1909 in a brochure: *The Hamburg City Park and the New Era* [*Der Hamburger Stadtpart und die Neuzeit*].

III

HOUSE AND GARDEN

A. Planning

House and garden don't only belong together externally. On a spiritual level as well, building culture is not to be separated from garden culture. One comes from the other. Interesting garden forms indeed have occasionally existed that were located at a distance from human living quarters, but never were there perfect gardens *without* highly developed building culture, nor the inverse. Indeed, I believe that it was someone no less than LICHTWARK who somewhere wrote that we had better not build country houses, indeed not build houses at all, before having useful, beautiful gardens once again.

Those who intend to erect their own house and garden must above all thoroughly and carefully consider the *first* measures to this end.

To begin with, the selection of the building plot is of decisive importance. The more or less protected *location*, the appropriate *size* for respective needs, the so-called *soil fertility, subsoil, water supply,* and *drainage* of all kinds, *streets,* plot *layout, sewage connection,* or *septic tank*—these are all questions that are critical to the future, harmonious thriving of house and garden, and for the unlimited enjoyment of country living. For example, I would never build my house on the north or east side of a hill, or indeed only if an existing view or superior vehicular approach justified this. On the other hand, I would welcome sterile soil in a poor area as a stimulating and difficult challenge. Just as obstacles are dealt with in life, so it is the same in the area of gardening where challenges inspire the most original garden creations.

Once the land has been secured then the question arises: Where is the house best situated? I have already discussed this in the case of garden cities: here the true gardener shudders, almost by professional obligation. Nine-tenths of all residences of the well-to-do today have their value reduced, or entirely negated, by the *improper location of the house* and its dependencies. The *building* should be situated to allow the garden to open to the south to the greatest degree possible, toward the sun. To be more precise: when there are no truly serious obstacles, one should place the house in the northwestern corner of the plot, so that the garden area as a whole lies open to the south or east. In this way, one complies at the same time with an equally fundamental requirement, that the main rooms of the house are oriented toward the south or east. And it concerns us little from today's standpoint, if as a result of these measures passersby occasionally glimpse service areas, while the house itself is removed within the depths of the garden. Only in respect to one's *own* habits can one comfortably and fully experience country life. And in this way a relatively perfect garden develops, that is, when its composition is *left open* to all the positive potentials of the site. When a house sits with a shallow area in front, with equally small areas on each side, and a limited back garden oriented to the north—literally "right in the middle" of the site, then even the most experienced gardener can do nothing sensible with the garden. It will always be just green decoration. These lapidary principles were first energetically formulated by HERMANN MUTHESIUS, whom I also thank for encouraging me in this work, and of course these are specifically valid for the prevailing standard types within our *Landhaus* settlements. Exceptional situations, for example, extensive land areas, significant differences in elevation, and good distant views, are always desirable advantages worth striving for.

The critical importance of correctly answering these initial questions on location should inherently lead to the application of all experience available today. Normally, the layperson thoughtlessly and submissively turns all this over to the architect. But that today's architects on average have long undervalued the worth of such considerations is shown by casual appearances when we stroll through the villa districts of our big cities. Finally, there are still building ordinances—counter to the essential desires of architects and those they advise, which would *not* come from them!—the so-called villa-clauses, with their provisions directly contradicting these and other requirements of the first planning phase. The well-being of individual residents as well as the future impression of the settlements are dependent upon the manner in which houses are related to garden and street.

As long as no definite improvement to these matters emerges, and yet no tradition exists, then conditions inherently demand that the presence of the garden designer is also necessary *before* anything of great importance has been determined. His experience, derived from the constant handling of soil and plants, renders him fundamentally capable of coadvising on settlement questions, at least in regard to the garden. Today's architect tends toward the view that just as with the layperson, the gardener knows little of such things and his decisions can only be a hindrance. But that is not the case. I know of a whole series of larger country houses, where owing to certain circumstances the layperson and gardener had to determine the position of the building *in advance of* the architect. No, I don't believe that practices that thus come about must be seen as absolutely detrimental to the magnitude and purity of the architect's conceptions for the advantageous position of rooms, the good interrelationship of living, service, and entertaining areas, or to the clever exploitation of distant views, even when judged from the highest standards. But such practices are of course uncommon. A good rule is: *owner, architect, and garden maker must start together from the outset,* if something positive is to come out of the project. The complex of questions is too large here, and only seldom will one of the interested parties *alone* be able to truly answer everything.

An existing *stand of trees* can also influence us in the positioning of the house. Particular rooms in the house, and its south side especially, benefit greatly from the intermittent half-shade afforded by tall trees. When these aren't present on the site, I often use a not insignificant portion of the available budget for some carefully chosen sturdy specimens of elm, poplar, chestnut, or linden of ten to thirty centimeters in diameter and larger to plant around the house. Today's technology is practically unbounded. Such sturdy trees with their compact masses of leaves are indeed priceless, as picturesque foreground for the garden composition and as a frame for the house itself.

Once the house has been assigned an appropriate position, then during the design process it should still be remembered that *the ground floor should stand no higher than thirty to eighty centimeters at most above the adjoining garden.* This measure is extremely important to allow for an intimate, practical interrelationship between house and garden. In England, where country living is known to flourish, one follows this in principle. No one should be burdened by having to overcome a series of ten steps and more to reach the garden. Experience has shown that the frequent use desired is not attained, sinking the garden to the level of a showpiece. As such it has absolutely no prospect of serving a cultural goal. The artificial *raising* of houses above level ground for so-called symbolic purposes that is still popular among us is an absurdity. This makes the garden unusable for social purposes. Those who don't have the means to erect their house on the land (literally[22]) as required by the particular conditions of the site, should stay in the city and dwell in high apartment towers. But when the site or situation demand a higher elevation for the house, then terraces should be laid out at least by the main rooms in order to provide a transition to the garden.

By contrast, *differences in elevation in the garden itself* are always welcome. They make it appear larger to the eye, bestow light and shadow effects, and intensify the spatial impression.

When it is not possible to bring the kitchen and service areas together within one building wing, then it often helps to locate a sunken service yard on the functional side of the house, thus making the cellar dry, airy, and bright; and there are more practical benefits here. There is no sound reason that seriously prevents us from rationally and practically orienting the house on an open site toward every direction.

But we see here yet again that country living is highly dependent upon the *very first preparations,* for the garden as well.

To this end, I also reckon on the securing of the *boundaries* against wind and weather, and in particular against people looking in. This really can't be too carefully treated. No one would consider building a house with walls filled with holes, for a hundred reasons. No one wants to be observed constantly in the most intimate state within their four corner stakes, not even by the most trusted neighbors. The house in fact contains such fine and tender, such personal things, *because* it is inaccessible to others. Thus, one may say that in the same manner raw garden land is only a real garden when its green material tightly defines the boundaries: *Gardens are spaces! What has been lacking in gardens until now, which is a prerequisite of true garden living in the new ones, is that they be enclosed.* Our treasure store of plants offers a thousand possibilities with hedges, blooming shrubbery, and trees for this purpose.

Just at this point, after the boundaries are secured, may one begin to lay out the garden with good conscience. From the beginning, the axes of the different rooms must be taken into account. As far as possible, the size, composition, and character of each garden section must be determined according to the use and importance of the room that it lies in front of. It would never occur to me, for example, to locate the flower or rose garden anywhere else than in front of the sunny living room, or the lady's dressing room, the evergreen garden nowhere else than in front of the north- or west-facing dining room, when nothing argues against it.

For the spirit of the contemporary garden, the domestic refinements of recent years, as expressed in the new country houses, are essential. A good garden simply *repeats* the customs of domestic living, *expands* them, and translates them into garden living. This is to be understood literally, to the extent that the typical rooms of the house should act as models for garden use. Is the comparison so far-fetched, if I occasionally refer to a central, tree-ringed lawn and play area as a "green hall," an open flower garden as a "garden boudoir," a wide terrace as a "healthy reception and communal space"? We have indeed always had a "kitchen garden," and what are the many pergolas in the garden if not "corridors of higher character"? Such green spatial compositions then are nothing other than the protection and framework for *dwelling* in the open in all possible variations: *the garden is our expanded dwelling.*

Seen in plain daylight, these roughly outlined principles on the relationship between house and garden would not be much more than truisms, if even the most modest of these was not still being incorrectly observed. There are yet more things regarding the planning of house and garden that could be mentioned to advantage. However, I will consciously limit myself here, and refer to the study of the given examples. What we should note is: *that the inner division of the house finds its respective organic continuation and translation in the garden, and that it echoes the characteristic formal units of the architecture within the garden.* This is the guarantee for a beautiful and practical combined effect, giving a greater unity of house and garden.

First and foremost, the subtle interplay of dwelling and green surroundings allows for such an intensified, relatively comfortable domestic culture, which justifiably appeals to us as healthy refinement that all classes should strive for.

B. Fundamental Design Principles

In order to make a good garden in the future, some of the old trappings must be left aside that today still count among the iron rules of "fine garden art."

Beautiful, old gardens usually leave us humans with a strong, spiritual impression. Old Adam, with his head in the sand, believed that he could cleverly take advantage of this situation. He had already inwardly engaged himself with this *beforehand*, and was convinced that afterward he would received a beautiful garden.[23] Many people today typically approach their garden plan with deep emotion, in much the same way. Thus, they make their task unnecessarily complicated. The kind of false sentimentality that is realized in the garden today is simply not to be believed. I don't know if this is indeed a lingering, particular feeling for the tender and "mysteriously living" quality of the materials that this field must work with, or merely the vestige of primitive mental processes: it is often a truly comic reality that even the most crooked land speculator believes that he must restrain a tear when he is about to invest the closely guarded profits of his activities in a garden home [*Gartenheim*]!

Romanticism is an old prop on the garden stage, and many good gardeners have made themselves wealthy through the artful, green exploitation of this minor human weakness. Nonetheless, I'm inclined to put the greater portion of the blame on it for the current misguided state of garden design and the reigning confusion in this field in general.

Making a good garden is in principle one of the most simple and matter-of-fact things: *one must organize it.* A garden organization will not be created, however, unless we are familiar with its factors and components, known through more or less established values. For the garden these are: the need for daily living in the open, games and sports, promenading and resting in green, direct use; and for values of a higher nature: the beautiful joy of flower, tree, and shrub, of cultivation and dabbling, of the rhythmic formations of forms and colors in the garden.

Whether high or low, these are always *defined purposes* that we want to fulfill through the furnishing of the garden, purposes that need to be expressed and formed. But how?

This question has been endlessly discussed over the past few years; thus, it is easier for me to be brief. We need only look around us, at how we are otherwise accustomed to satisfying the practical requirements of daily life. Such as with utensils. A spoon is naively shaped for "eating"; it is "fit for purpose," to use a term often used today. And the appearance of a chair, for example, is always only modifiable

PLATE 7

within set principles. Its ancient form has been fixed for thousands of years, and so it will remain, as long as people don't find reasons and opportunities—to newly orient it by looking to the past.

Just in this way, a long, long time ago, people also invented the *garden archetype* most suitable to their nature. This is the geometric. It is cut to fit "gardening." The origins and history of the utilitarian garden always showed the same basic geometric forms even among the most different peoples and climates, proving the correctness of using these essential human constructive elements for garden forms. Following these first attempts at intensive field culture by settled nomads, all of the higher garden types of subsequent eras were developed with progressive certainty, wealth, and refinement, including the "natural," which only as the result of decadent cultural conditions were natural.

Thus, we will attempt to consider the contemporary garden in light of this inclusive term, the *useful.* The garden must once again be naive if we hope to seriously develop it culturally. Plainness and naturalness within the conception of all newly formulated things is definitely, I believe, healthy, fertile ground for everything that our external culture should bestow on us. For the formation of our garden, we must begin organically with the simplest concerns, reserving all the potential refinements of later development. *Simplicity* is the best capital investment for the future of our garden intentions. Rich and splendid gardens are not always beautiful, and moreover today are primarily put together using symbols of gardens from earlier periods. Simple, naive gardens are like little children, which for the moment still don't mean much to the world, but which are later its hope. Simplicity and usefulness go hand in hand with a healthier, more prudent cost outlay, with an *economic* sensibility, which should never be lacking when basic needs are being considered, such as gardens.

Just on the basis of these three fundamental principles for today's typical garden, the necessity for the geometric line in the garden emerges a hundred times over. There is no form of the so-called naturalistic garden in which the terms simplicity, usefulness, and economy can be exhaustively applied.

However, the application of the geometric line as the first spontaneous and most important human design element, not only in the garden, still has not been exhausted despite its thousandfold adaptability. Initially this first line facilitated the expansion of the garden in *length* and *breadth,* then together with the improvement of domestic life and its free transfer onto the garden, the third dimension was realized through *height* or *depth,* as of itself. Garden compositions arise with cubic forms, that is, of true *architectural appearance*—each of the previously discussed

plate 7

A garden house in the middle of a flower garden; the large windows that can be lowered open the view to three sides.

garden spaces arises. And in this way the higher level of garden conception is reached, which in the end is always concluded by recording the overall appearance of a people's strength and refinement through monumental garden documents.

This example is also valid for the counterclaim. Everywhere within our current life organization we must reckon on organizational possibilities if we want to keep our feet on firm ground. Therefore it's inconceivable, as seen from any side, that the garden of the poor just as the garden of the rich, or that which belongs to both—in short, it is technically and economically inconceivable that we would want to construct the just and good garden of the common man with other means than the simple and universally understandable form of the geometric line. Otherwise, irresponsible fantasizing will result. But one may turn this argument on its head: the contemporary garden must *therefore* present a regular tectonic appearance, because its own, new social and economic character can only be fully utilized in *this* way. Consequently, the transfer of architectural principles onto the garden is to me not merely a formal and not even an aesthetic matter. My garden doesn't look architectonic *because* other earlier garden epochs used the same means of designing, and people love change, and not only because the regularity and control of the geometric line promises a stronger, more rhythmic effect in opposition to the arbitrariness of the free—no, the architectural design of the garden is necessary for us because it is so *simple*. Because its elements are easiest to handle and are economical by nature, so that in our era of mass-problems *it alone* results in an effect on a broad scale: *I want the architectural garden for economic, social, and ethical reasons.*

Therefore, the question so often put forth today, of the fundamental design of gardens that are not directly influenced by architecture, has little justification. Even larger house gardens, the park, and wide open public gardens of all kinds are underpinned by the primal architectural law of all human feelings and pursuits, a law that, although developed from the house, neither originated with it nor was defined by it alone. From these grounds it logically follows that if we want a new garden culture based on our *own* life, then all of its essential expressions must serve the *architectonic gesture*. The "architectonic" is a fully free expression so diverse in itself, especially when applied to the garden, yet only philistines and the small-minded don't accept this, for they are hypnotized by external qualities of "constraint and rigidity." Just how bearable this "constraint" is in truth, I can no better illustrate than through the evidence of the accompanying photographs, which intentionally show the most rigorously designed architectural gardens.

Thus, today we want to see definite functions fulfilled following a definite, preconceived rhythm, just as with the house. Therefore, it is unavoidably necessary when someone plans a garden that he first decide what he really wants, that is to say, that he clearly express his thoughts and needs. I admit that when confronted by a garden plan without knowing anything about the client—whether an individual or the mass public, it's the same—nor knowing very much about the kind of measures noted above, I always stand by quite helplessly. I don't place much value on garden projects that are pure fantasy. Thus, my first question is always: How do you want to use your garden? Do you perhaps need a ball field, a playground or tennis court, or perhaps a sunbathing area, a swimming pool, or a covered pavilion? Do you like the sun, or more shade, to walk about, or to gaze quietly, and do you have many guests? Should orchards and vegetable gardens cover the needs of the household, and which varieties do you prefer? Oh yes, and what about the planting in general? Do you particularly like red maples, magnolias, rhododendron, and does the lady of the house have a penchant for the various "charming" perennial flowers—very good. Naturally, roses must not be forgotten, and please, a laburnum shrub as well, and perhaps some fine lilacs, of red and white. Finally, how do you feel about adding some artistic ornaments? Is a colored ceramic fountain, a statue, or an ornamental vase desirable? In short, there can't be too many questions, for the whole story often borders on indiscretion.

Being thus armed allows for the creation of something excellent. And everything depends on that alone.

C. Composition

Just how necessary such overall harmony is, I will show through the following three examples taken from my practice, which demonstrate this much better than theories alone could.

With the first (the W[egmann]. Garden, in *Rhede, W[estphalia].*), the architecture, which has Hermann Muthesius as its father, facilitated the creation of a good garden from the beginning. The siting of the house at about two-thirds the length of the rectangle, open to the south and east, with service spaces placed to the north, is beneficial for the division of the garden. The main living floor is raised several meters above the garden by utilizing a rise in elevation that crosses the site, connected with the entrance by a path on a slope. The lower floor, supported in part by arches, leads directly to the garden without any steps. The driveway lies on a terrace ornamented by a pergola, further emphasizing the change in levels.

Through this insightful preparatory work, the most important things had indeed been done for the garden at the same time. For me it was only left to functionally lay out the remaining areas for the varied requirements, to interweave appropriate vegetation within, and in addition, to consider how to effectively reinforce the especially rhythmic composition of the house through the garden.

Thus, the entire front section of the garden was reserved for a *communal lawn.* This serves children's play, adult sports, and social gatherings at festive times. As a surrounding promenade, a hornbeam pergola protecting against sun and dust accompanies it as the boundary to the street. This hornbeam path is open toward the interior through arches, and continues next to the house by forming a small

plate 8

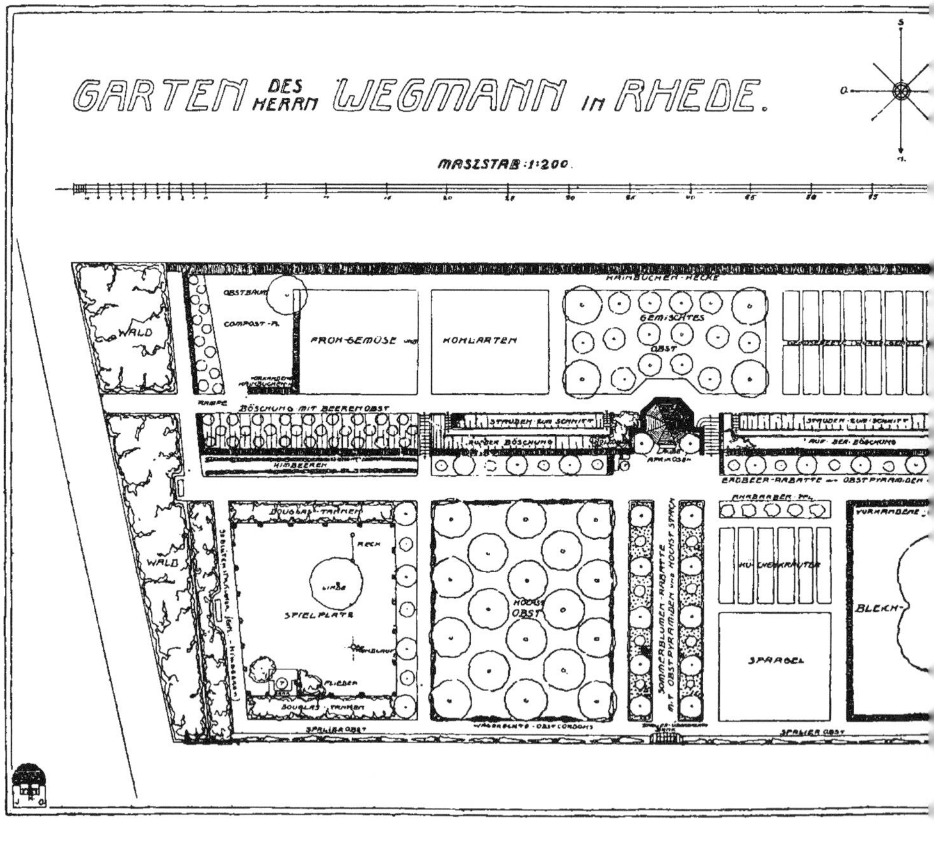
GARTEN DES HERRN WEGMANN IN RHEDE.
MASZSTAB: 1:200.
WALD
COMPOST-PL.
FRÜH-GEMÜSE UND KOHLARTEN
HAINBUCHEN-HECKE
GEMISCHTES OBST
BÖSCHUNG MIT BEERENOBST
HIMBEEREN
STAUDEN ZUM SCHNITT
ERDBEER-RABATTE UND OBSTPYRAMIDEN
WALD
DOUGLAS-TANNEN
SPIELPLATZ
LINDE
RECK
FLIEDER
HOCHST. OBST
SOMMERBLUMEN-RABATTE M. OBSTPYRAMIDEN
RHABARBER
KÜCHENKRÄUTER
SPARGEL
BLEICH-
SPALIEROBST

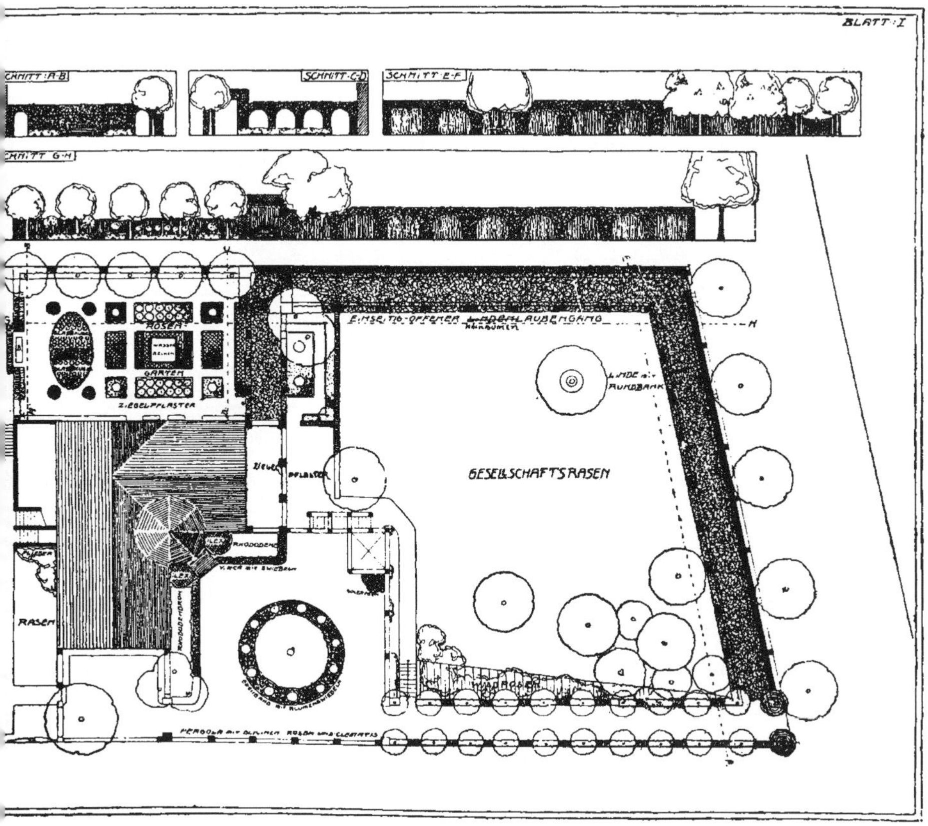

plate 8

The Wegmann Garden in Rhede, with plan.

arcade of four arches. The remaining open side is completed by a birch allée, with a wild-rose–covered embankment in view, while freestanding trees, chestnuts, maples, and beeches are distributed as needed for shade, planted as striking specimens that break up the strict symmetry.—One should observe how the inherent freedom of the garden-architectonic is manifested here, and how this is based on the attractive result derived from these purely functional, yet at the same time higher rhythmic intentions in these few simple measures, and just how few artificial accents are needed to bring this all about. This is all self-evident through the overall division and self-evident in the vegetal cladding.

The garden design, as developed up to this point, also advantageously includes a *terrace*. It is designed as an evergreen and spring garden, and from its shady corner seat one has a beautiful view of the green panorama at one's feet.

On the other side toward the south, a *rose garden* is located next to the house. Situated on the same level as the front garden, the play of flower colors cheerfully peers through the open arches of the pergola. Within its flower beds roses and lilies are carefully ordered according to color and blooming season. Around a central basin, blue water forget-me-nots and white narcissi bloom, and even a border of sweet-smelling herbs isn't lacking. Here the garden lover may stroll about, sheltered by high hedges and red flowering chestnuts.

All three garden sections, each in its own way, expresses the sharply defined, varied needs at hand. Since they are used by the same people at the same time, it was an important task to fuse the parts into a unified garden structure, conveniently and organically connecting the parts with one another. These garden sections were reserved for the undisturbed usage of the family and their guests; it becomes a *communal garden*, not too big, but intimate and full of variation in its design, edifying through beautiful images.

Clearly separated from these garden sections, an extensive *utilitarian area* stretches out behind the house. It contains everything that the heart of a practical housewife could desire. Every available corner is put to use. Beginning with a little herb garden and the trellis on the south border, to a small asparagus planting, which, however, isn't in a favorable position. It was already there, and should be removed later, and the area it now occupies next to the kitchen garden was connected to the drying yard. The sunny embankment was transformed into a *cut-flower border*, which has to serve the needs of the household throughout the year. It is divided by an elevated pavilion, from which one enjoys a good view. All the way to the back lies an outdoor gym for children, protected by fir trees and old hedges, which also includes a few planting beds.

That was the general description of a *Landhaus* of typical local character and overall area (ca. five thousand square meters), which embodies the potential for a well-considered unity of house and garden, and for a higher expression of country life through the simplest means possible.

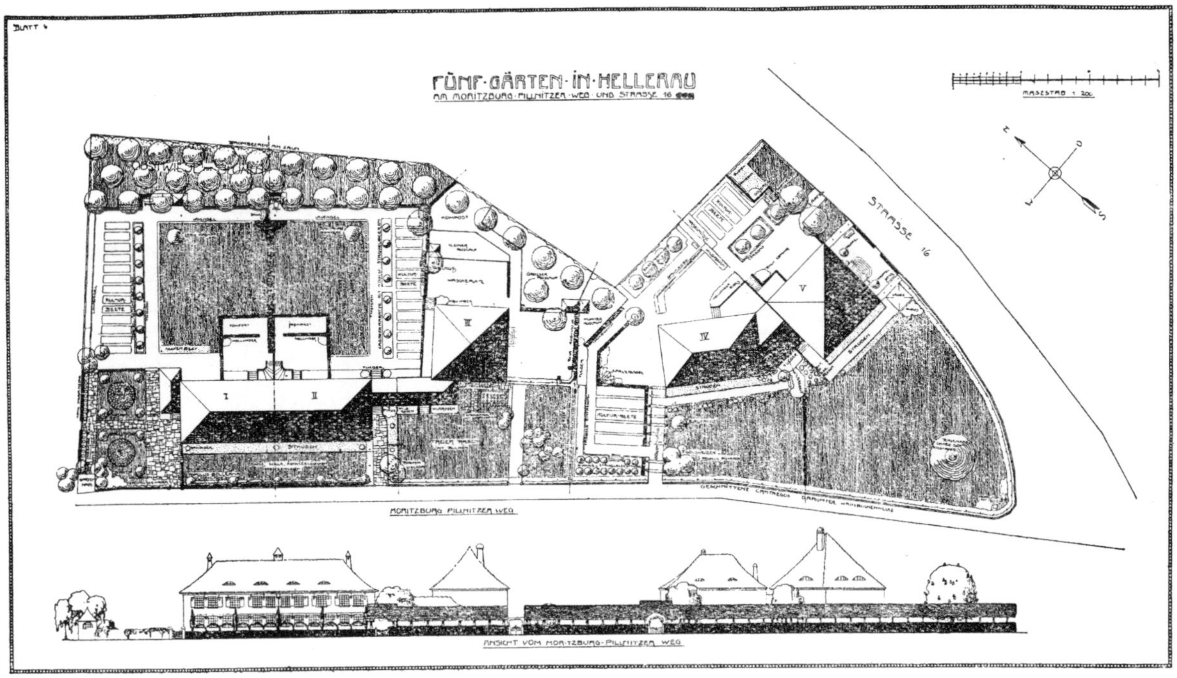

Five gardens in Hellerau
L. Migge

TRANSLATOR'S NOTE

The garden city of Hellerau (1909), near Dresden, was one of the first in Germany; it was closely associated with the German Werkbund and Hermann Muthesius. See Clemens Galonska and Frank Elstner, *Gartenstadt Hellerau / Garden City of Hellerau* (Chemnitz: Palisander Verlag, 2007).

At least this garden may be considered comfortable, in contrast to the five *small house gardens* that I laid out in the garden city of Hellerau.[24] The whole land area is no larger than about six thousand square meters. After deducting the building areas, each garden is left with just under one thousand square meters on average. That means economizing, even when in principle the owner of such a small plot scarcely has fewer desires for his garden than his more well-off fellow citizen above. Often to the contrary. Here the means must be found to bring more of the functions together in one place. Thus the larger *front area* by the house entry is likewise a *children's playground*, the *ornamental lawn* is also a *drying yard*, and the ever-present *service area* is combined with a *compost facility*. As a rule, everything bears the stamp of the most rational utilization. All blank facade areas and *fences* are covered with blackberries or espalier. As "ornamental shrubbery," wild roses and elderberries are planted in suitable spots, and currant shrubs form enclosing hedges. Of course conditions here were not as generous as in the previous example.

These houses designed by Prof. Riemerschmid are favorably disposed on the interior and exterior, but in regard to the garden layout their position on the site is less so. For indeed, as we know, it is always a problem when the secondary rooms of the house that lie in shade front on the main section of the garden. In spite of this, everything was well integrated. It was particularly important to realize that these houses sit on a dusty road, and that they were only partly lived in, with the other part acting as a standing exhibition of the interior. Therefore, the kind of green enclosure that would protect against wind and weather, create green spaces, and offer only limited inward views to passersby had to be carefully taken into account. In addition, because in this situation pleasing formal expression was not a secondary matter, and because the means available were extremely limited—I have always felt that: the garden is ultimately a calculation. It is a compromise between local conditions, numerous desires, and available capital, a combination of ideal intentions and earthly limitations.

Our third example in a sense embraces the occupants of both of the previously discussed categories of garden types, and their life and their desires to be in the open. This is a public garden, a so-called play park, the first one in Germany at that, and it is only included here because of its significance as the natural extension of the private garden, as an illustration of modern garden design principles.

There I found a generally level meadow of about one hectare (ten thousand square meters) in size, and about one meter higher than the surrounding streets, edged on one side by a high hawthorn hedge. Everything was left as it was and put to practical use. The *sunny meadow* was treated as the core and most essential element of the site. On this plain dotted with dandelions, bellflowers, clover, and colorful grasses, the adults are to find rest, and the youth to indulge in games, sports, and spirited horseplay; footpaths cross it, in the foreground a chestnut throws deep shadows over the inviting round bench. This meadow is bordered on two sides by a *linden pergola*, with a dense leaf cover affording cool shade, while its arcade-like open sides allow for open views (which in any case is similar to the typical growth pattern of lindens). It should invite promenading.

To the north, the aforementioned *hawthorn shrubbery* is separated from the meadow by a path, providing shelter for all kinds of wild flowers and perennials that multiply by self-sowing, such as marigolds, poppies, finer varieties of clover and bellflowers, and others sprouting up in masses and giving this path the character of a *flower path*.

The last side of the meadow is enclosed by two shady *groves* that serve as shelters from sun and weather. They are distinguished from one another in plan, but harmonize through the selection of varieties of red maples and birches. The tree masses of these groves hold between them a small *playground and outdoor gym*, well furnished with all kinds of play equipment and a large sandbox. The background is formed by a row of striking pyramidal poplars, with *benches* set in niche-like spaces in between, for onlooking mothers and maids. A sunken *flower garden* is located in front of the small playground. This is conceived as a perennial and summer garden with many flower varieties, at the same time serving the need for beauty and instruction through its content and organization.

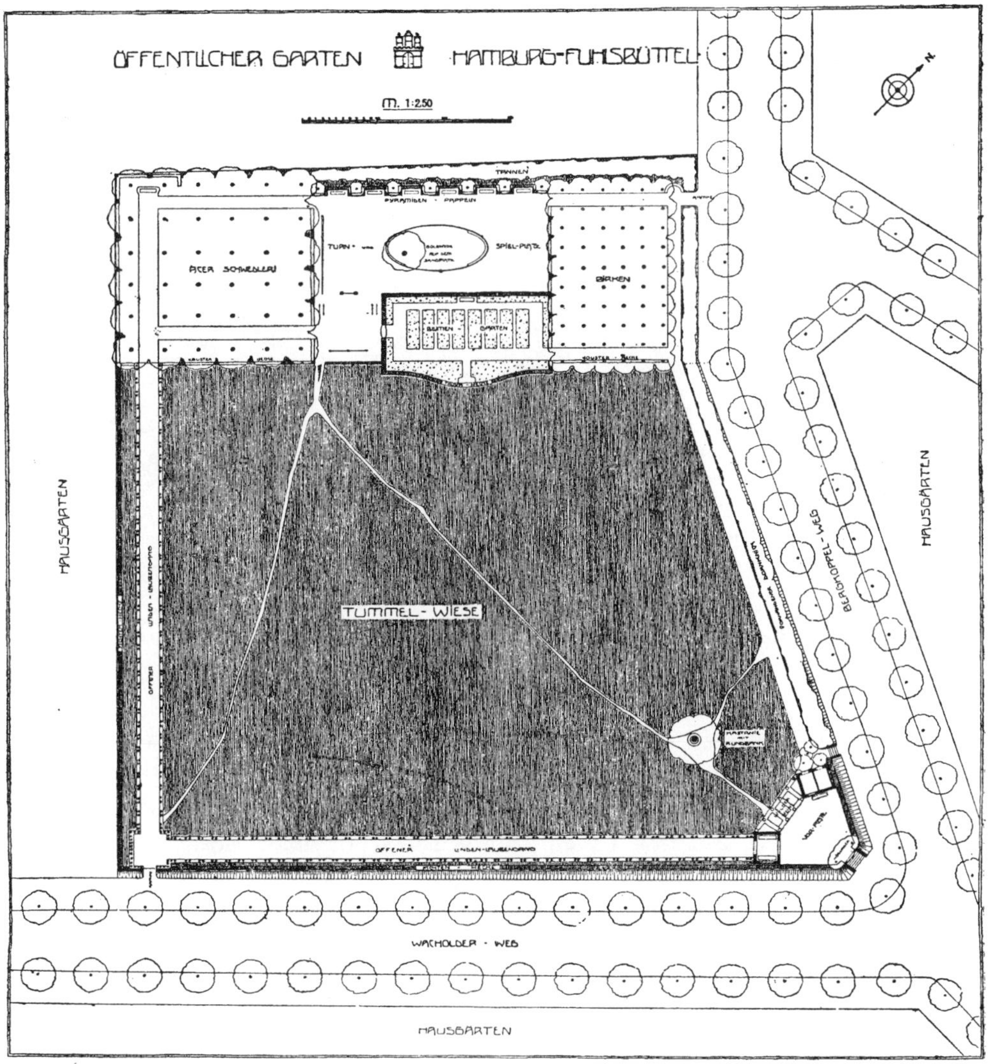

Plan of the public garden in Hamburg-Fuhlsbüttel
L. Migge

TRANSLATOR'S NOTE
This park has been recently restored.

Just as the hawthorn hedge to the north gives protection and natural definition to the whole site, a similar one was created on the west side in the form of a high spruce planting to block the prevailing winds here. Low spruce hedges bound the garden on the remaining two sides.

The site is accessed through the *main* and *secondary entrances*. At the former, red brick steps lead from the street intersection through an elderberry archway to a forecourt. This entry court leads toward the garden to two simple little houses with a *shelter* and a *public toilet*, connected to each other by garlands of vines. From this point, one may easily reach all parts of the garden by the shortest paths. The secondary entrances to the south and north have ramps for baby carriages, and the neighbors have private gates.

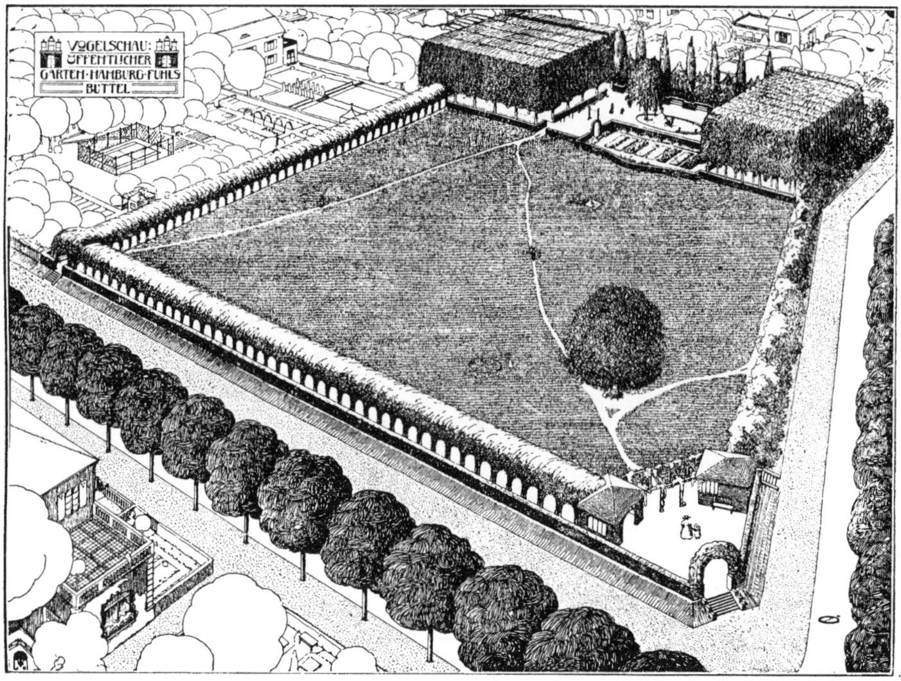

Aerial view, public garden Hamburg-Fuhlsbüttel
L. Migge

That was a full description, and admittedly it sounds very orderly. But in reality it's not so rigid. The individual forms are ordered following a will to an overall rhythm of a garden-architectonic nature. This is not conceived as being mawkish or sentimental, nor as picturesque-romantic, but as a human product resulting from the self-confident, worthy order of human purposes.[v]

From our three examples it now becomes ever clearer that their form doubtlessly had to be regular. How was that then?—out of *simplicity*, out of *economy*, and out of *beauty*. Then even the most stubborn opponents of such garden forms must admit that the number of facilities that I've planned for our public garden cannot be matched through any kind of naturalistic garden form, when given approximately the same space and the same relatively small expense. Here only genuine objectivity can reach the goals. What has been promised here, the clear, refined appearance of functional forms, is also served to a large extent by the highest goals: harmony and beauty.

v It is all the more painful, when the effort and sacrifice that was required to create this garden under difficult conditions at the time are now ignored by the Hamburg city engineering department through their disinterested maintenance of the new installation.

IV

PEOPLE AND THE GARDEN

I count three types of people as being of particular relevance for today's garden operations: *garden experts*, *architects*, and *laypeople*. Each of these tribes has already been of great service to the new garden. However, as a whole, we indeed often find among them hard egoists, shortsighted specialists, and ignoramuses of the worst sort. How can this be improved?

A. The Expert

We will start with the *expert*. One should respect him—but not idolize him; as things stand, the expert is a bugbear, unshakably rooted in our life. The expert in his purest form is an embodied convention, and therefore culturally sterile as a result of professionalism. The contemporary professional is the more or less upright, born supporter of the current world order, and on a national level he represents the reliable average man. Here we think of the "petit bourgeois" in the usual sense of the term. Almost all great acts within specific fields were realized by non-experts, as a matter of honor. The specialization of the professions was certainly brought about by the victory march of "modern civilization," but does one want that in the future the world will be entrusted to the expert *alone*, as it sometimes appears, in which case I'm certain it would quickly become insipid and rather barren. We must protect ourselves from the "expert"!

And the garden expert? He is just the same, who even today struggles with desperate energy so that the linden trunks in an allée are perfectly upright, and that each shrub is exactly a meter from the next, that "the" fruit tree is trimmed into a half-sphere, and that no daisies spoil the lawn, etc., etc. In spite of all this, he is highly overestimated, as with all experts in general.

First comes the MUNICIPAL GARDENER. He is at the natural peak of the gardening career path. Indeed, every official represents the highest point of human existence for the dear German public. And so one thinks here of the ruler of the gardens of a metropolis. He is a little potentate. One literally honors him—in

PLATE 9

the sight of those wonderful green promenades, and many believe him to be in a superhuman alliance with the powers of nature. Well, it's not that bad. Those who know how such things really come about, and that the dear Lord does them anyway, think much more skeptically about the creative power of the gardening bureaucracy up to now.

Through a conservative inheritance from an earlier time and through clever means of assimilation, gardening officialdom undoubtedly has much more influence at the moment than it deserves. They control the press, mold the younger generation, and much more. This is not without peril, no, this is *sufficient grounds alone* to utter these words here. By cleverly exploiting this situation, in addition they have achieved the status among the broader public of being the real representatives of the renewal of German gardening. When somewhere today the discussion turns to new garden design, the Smiths fixate on the green urban square in front of their windows. And above all, it is the self-satisfied mannerisms of these people in particular that is annoying.

But in reality, things are just the reverse. It was highly educated, courageous laypeople who pointed toward new directions for the garden. Then came Behrens, Läuger, Olbrich, Muthesius, Bauer, Hömann,[25] and others, free artists and young gardeners, who laid on a fresh hand and set the new ideas into motion. *They alone are the conquistadors of the new garden territory.* New garden firms were founded, and the old ones soon decided to change saddles. A bold wager in fact, of fertile studying and testing, and I know more than one free soul who six or eight years ago stood at a tipping point—only because they obstinately fought with holy conviction and their whole being against a pitiless phalanx of counter-interests, and the likings of Philistines and Pharisees. *They assumed the financial risk.* This deservedly must all be brought to light.

In the meantime, during all these critical years, officialdom stood almost unanimously on the side of the reaction. One reads this in the press. And now? A journey through our beautiful German cities, aside from a few such as Charlottenburg, Essen, Cologne, Lübeck, and others, reveals with sad conviction that even today the official new garden rhythm is poorly provided for, even now when indeed there is nothing more to risk!

Every bureaucracy has its defined and important tasks of a more organizational nature, thus possessing the special talent for this. But when it tries to hinder the productive energies of a progressive people through direct resistance or by clever passive means, then it must be quietly put in its proper place. Before the field of new garden design can develop in a generous and free manner for the

plate 9

View from the position in Plate 1 seen from the other direction. The bridge is intentionally elevated to allow for boating and ice skating.

important goals of the overall culture of the German people, it must first be freed from bureaucratic red tape.

In recent years, there was a forceful, elemental tendency toward new divisions of labor within many professions: more than in the past, tasks of a creative kind should be reserved for free competition among those who are qualified and professionally active in this respect. Architects first led the way down this path with success. Even more than the area of house building, did that of the garden demand development in this direction. Beginning with the dependent qualities of our special materials, that require being treated with ever newer freedom.[26] The first and by far the majority of the new fundamental experiments in modern urban gardens were carried out by independent artists and garden architects. German cities would be better represented within new garden design if city fathers would decide to hand over their problems to experienced innovators in the field. At least in the beginning. This is particularly valid for *smaller cities,* which of course can't afford outstanding permanent staff because of costs.

Minor resistance to unreasonably aggressive bureaucracy thus also belongs to the chapter on the freedom of the garden. However, in this book dedicated to an objective garden enlightenment, I won't fail to touch upon the shortcomings of the other side as well, namely, those of the GARDEN ARCHITECTS. The fact that their deficiencies lie mostly on the *economic* side doesn't really affect the essential superiority of private garden architects in all *productive matters* of the garden, as I consider them in reality.

If the factor of ability alone once sufficed to solve cultural problems, now on the whole it suffices no longer. Today the realization and propagation of new opportunities resulting from a different *economic form* plays an essential role. Much of what could earlier be accomplished through personality, material, and quality alone, is now being taken over by commercial organization. Nine-tenths or more of all gardens are now designed by commercially active gardeners, and it is immediately clear that the means of production under these circumstances must have a strong influence on the final value of these garden products: *real garden culture for us is no longer imaginable without the confident contribution of the "businessman."*

But there are problems with this kind of support. It must be concluded that at the moment anarchy more or less reigns within the economic side of garden life.

The strong new garden movement and the intensified field of work that goes with it have brought about an entire range of large-scale operations whose directors, with few exceptions, are still lacking the necessary sense of ethical responsibility for their work. There, a business-like "machine" is set up, a naturally "first-class" artist is engaged, and the whole thing is ready. Here, they often have little clarity concerning the most basic conceptual and practical requirements of the modern garden, but they glibly mime their way through it. A little bit of green and flowers in contrast to a piece of trelliswork or a white bench, straight paths, some steps, pavilions and pergolas, often with impossible structures—and the "contemporary garden design" is done. It doesn't matter how: just *do, do, do*; that appears to be the solution!

The same impression naturally applies to the people as well. Today, the so amusingly, variously named landscape gardener, garden artist, art gardener, garden engineer, garden architect, and whatever else, here in trusted collaboration with the garden superintendent, director, inspector, master, etc., as he may call himself—he wants above all else to be an *artist*. I don't know why this is necessary since there are so many other good and useful kinds of people. But there is a special gardening jargon that teems with superlatives: "artistic design," "refined concepts," "generosity," "ingenious creation"; these and similar judgments are proclaimed when they want to defend their right to exist through a modicum of useful daily labor. Thus, this species has yet to attain the status of the sincere, congenial professional man.

These are the little people, whom I assume carry out their work in similar ways in other professions. Nevertheless, for a transitional period like ours they are significant enough to be able to be taken seriously. Our life today is based on such complicated factors, that in the case of every kind of practical work such as everyone must carry out, a clear perspective is required everywhere. Self-congratulation and romantic eccentricity is ill suited to support this kind of life, and odd-job work of all kinds completely drives it into the ground.

In order to avoid this in the garden, we experts must above all decide to set aside the *banners of artistic gardening* in the corner for the time being. One therefore should think more on the homespun, but thus more useful value of dignified workmanship. Through the simple understanding of his productive work, through the soundness of all his measures and the solidity of the business he runs, the independent professional can contribute much to the curing of current conditions in the garden. One may readily remark that this is only achieved when thoroughly carried out. Don't look so crestfallen! I have long renounced the title "garden *artist*," yet I was still able to create an attractive series of gardens that brought joy to their owners. In future, if a garden maker stands before his client with the question: What are your needs?, then he is capable of fulfilling this with much expert knowledge, flexibility, and accuracy, thus achieving a great deal; on average, he cannot, and on average need not, manage to do much more.

Above all, such a simplified, almost natural treatment of garden work will result in well-crafted forms in the end, just as has always been the case with the higher forms of craft and applied art, and just as we vitally need for the greater number of small tasks in garden work. Here and in the areas of front gardens and smaller house gardens, the domain of the "freelancer," the indecisiveness of today's garden thinking has had a particularly devastating effect. The small landscape gardener, "confused by the favor and the hate of parties,"[27] and made even more

PLATE 10

helpless by increasing business competition, is in the habit of plunging at his garden victim with spades, shears, and a truly berserk fury. Today, he is inspired by the "architectural," tomorrow burdened by "landscape tradition," and if possible, he also wants to "express himself artistically." Thus, he ruins the external appearance of our cities. The only thing that will help here is the intensive dissemination of a modest garden idea based on objective principles, which through its application will bring about simplified workmanship.

THE SUPPLIER This garden expert, in the broadest sense, can also contribute to a generally valid overall clarity. Because within garden activity as a whole, as we see it, the behavior of those who *produce* and *supply* plants is not unimportant. Here it must be said: the average German nurseryman and flower grower today works without the least inner sense of the rhythmic-spiritual, indeed still not even occasionally with the economic-qualitative processes around us. And there is absolutely no question of any kind of more intimate relationship with the lay public.

Nonetheless, our supplier believes that his livelihood is threatened by the "modern," and he protests (even if these days only inwardly). I have met a whole range of respected experts from this branch of gardening, who still today consider our efforts to be fashion tendencies: "Another kind of art nouveau—it'll soon pass." In keeping with this, they make no arrangements whatsoever to adapt their cultivation practices to new needs.

Indeed, the ambition of every one-acre nurseryman to make a botanical garden on his little strip of land is an economic absurdity. He cultivates a thousand varieties—and naturally isn't really knowledgeable about *any* of them. Gardening business operations would develop much more smoothly if our nurserymen could specialize more. The names of each lily and clematis firm, and those from whom one can easily obtain transplantable trees, are easily recognized by the public, but the general supplier is Hecuba to them. Not least, it was the clumsy manner in which German plant dealers were retained by many garden architects to set up their own nurseries for them, so that they could obtain without difficulties the plant materials that they believed to be indispensable to their designs.

The advertising materials of these supply gardeners are also largely mired in old habits. One may take any catalog off the shelf. Modifiers such as *beautiful*, *especially* beautiful, and *extra* beautiful grace every page. This ridiculous fanfare blooms in its special luxuriance, and it receives a worthy crown through the infamous "selection of our own choice." But one may well infer *that*, namely, the "noble selection," "giant selection," "elite selection," as grandiose as possible—repeated a

plate 10

Boot moorage and swimming area with steps and a flag mast create an harmonious picture.

hundredfold, must be brought down to the level of the day-to-day public. Or, one may leaf through a rose catalog (readily from the most famous firms), and note, as I now do on a random page "x" from the top: *cherry red, bright pomegranate red, blazing* red, *vivid* red, *tender flesh pink, cinnabar red, carmine, silky* pink, *fiery dark red, lustrous* red, and the roses there are even redder. Twenty pages later, only someone who is colorblind will notice any difference. I ask: Why didn't our otherwise practical gardening class long ago possess official color tables, that at least would inform their not yet fully authoritative eye? Fruit is also not quite logically categorized: *early* and *abundant yield*, abundant yield, *very* abundant, with *greater* yield, etc. I read one after the other of these among the apple labels of a large firm. In the same way, the naming of cultivars such as, *"Triumph," "Ideal," "Fame," "Emperor and King,"* all in all the garden-related coinage of some kind of accessible language that the proprietor tries to be at the forefront of, but certainly not within the realm of good decorum. From there on to the latest nouvelle humbug where the highest art of language is at home: "*Dahlia, white lady* is doubtlessly the whitest dahlia of the present." Poor Boieldieu![28] Or: "*Shriving's improved yellow English gray-red-skinned giant-turnip or rutabaga*."—Ugh, what's the tuber called?

They know not, our plant cultivators and handlers, how foreign these extravagant, dreadful catalog entries are to the public; when buying, the expert in fact primarily uses the semiannual autumn and spring price lists and classified advertisements. But *that* is it indeed, for as traders, each believes that they don't need to be objective. This is clearly perceived from the layout of these price lists. There aren't many businesses anymore in Germany that would dare to amass so much tastelessness within such limited scope. But the symbol of Flora strewing flowers in a kitschy, brightly colored reform dress[29] as painted on a cover, appears to our world-class horticultural firms to continue to suffice as the sensible expression of their capabilities and their future.

However, there are also a few practical sales books with well-organized content, such as that of the Erfurt seed firm, *Haage & Schmidt*,[30] which stand alone in the German catalog forest; with their painstaking diligence over the most modest entries, this is in itself an achievement for garden culture. At least here one also sees signs of an improvement. An impetus to this recovery has also been provided by the objective and tasteful advertising publications of a few independent garden architects at the same time.

All in all, one sees that: even in one's own house there is enough to do. We all have much to improve. The whole process of recovery would be greatly assisted through the early training of young gardeners. This must be introduced in the technical schools. The highest level of gardening instruction, along with the necessary technical knowledge, must minimally provide a certain feeling for proportion, a small measure of taste, and in the first instance a general understanding for the overall cultural and economic events of our day. This is the opposite of what is taught in our schools today under the heading "garden art," which could largely be omitted to advantage. The modern perception of form found in today's applied arts schools should also be made available to young gardeners on sites with living plants. Everywhere, garden design or horticulture classes should be set up, and much more besides.

Thus over time, we could of course attain unified, great garden conceptions and their expression, such as has distinguished all monumental periods. But this development is unimaginable without serious, practical preparatory work in the garden. Yes, *our gardening practice will really only begin to be culture when its economic conditions are first reconciled.*

B. The Architect

But there is yet another man who asserts a claim on the new garden. That is the architect.

From the viewpoint of the modern architect, there are three particular grounds that support his interest in the garden. The advantage of having the house and garden *unified* under one hand, the assumption that the appropriately trained architect has a *spatial* sensitivity, and finally while people are people, those who are knowledgeable—are 10 percent.

The Renaissance is often referred to for evidence of the natural logic of the first of these principles. Its particular circumstances often permitted a comprehensive task of the broadest scope to be placed in one pair of hands. Occasionally the house and the garden as well. As a rule, the "engineer of waterworks and artistic gardener" was placed under the direction of the architect. Under the French kings, who arguably created the most monumental gardens known to cultural history, the garden architect Le Nôtre worked without restrictions alongside Mansard, while in Germany Pigage[31] created the engaging garden rhythms of Schwetzingen free from architectural supervision. There really were not very many of these precedent cases. On the whole, they were also not really the standard. We now live in an age that is absolutely not parallel to any other epoch in terms of its method of operating. Our materials and the way that they are handled have multiplied so enormously, that this of itself has led to a minute division and specialization of labor. The "comprehensive artist" lost, once he had fulfilled his important role as the instigator of essential principles, but he gains in reputation the more we delve into the essence of new things, and recognize the necessity of their fundamental treatment. This observation argues for an almost effortless mastery of the respective materials, particularly for our aggressively accelerated methods of production. But in this connection, the higher powers have provided for the garden in such a liberal manner that it indeed always remains on the level of the most frivolous and damaging dilettantism, that is, if architects *as such* approach the garden. It is just these efforts of the creative talents among them, of whom I know a large number, which strikingly illustrate the statement: that *idea* and *spatial feeling* are not enough alone to make a garden. There are valid exceptions: but admittedly the majority confirms the rule.

Many architects have recognized this search for other means of contributing. They design the structure of the garden, and then let just any ordinary gardener fill it out with plants. Thus they usually just go from bad to worse. Aside from the fact that one may no longer speak of a unifying hand in this case, in addition the usual result is that the "landscape gardener" who is completely unfamiliar with the profession of the architect ruins the remaining positive *architectonic* concepts through incorrect plant selection and inadequate execution. Thus, the architect who created it becomes the one who ruins it out of superficiality or ignorance.

Final results are not prevented by such treatment. Throughout the year, I have an attractive range of gardens to redesign, which were started by architects. And for me it is always a particularly favored method of success, to lead an indecisive client to this or that garden designed by an architect: "See how these Taxus niches with their lovely white benches are formally well conceived. Only unfortunately the designer didn't take into account that these yews are crowded out by competition under the cover of these old elms, and won't thrive in the long term, and his cut-rate advisor didn't take appropriate precautions against destructive tree roots. Now this artwork looks rather sad."—In most cases, the psychological reflex to this small talk results in a contract for me.

Whether acknowledged or not: all of the *obviously misconceived* garden attempts of architects in fact belong under this heading. Thus in an unfamiliar profession, the interloper can only realize either the *unique*—or *nothing at all*. The midrange is covered by the experts themselves. On the occasion of urban design competitions and other events, I have seen garden projects by famous architects recommended that were childish in this respect.

Under these circumstances we cannot progress any further. In the long term, this must lead to a demoralization of the garden. Either the architect proves his superiority within our garden culture, in which case we experts will willingly, even gladly follow his command: the *capable* must lead under all circumstances! Or we assume that modern architecture has its hands full with its own shortcomings. Then it should objectively and gladly employ the respective trades, similarly to how they have long been accustomed to call upon sculpture and painting. These legitimate deficiencies will definitely not persist today for very much longer. It is certain now that a capable gardener trained in the latest methods is able to observe the architect's highest intentions through his own garden designs—without neglecting the vegetation, as others have done.

C. The Layperson

But even more important than the conduct of these two professional categories, the garden experts and the architects, is that of the layperson, who can bring about the healthy development of new garden living. This topic has hardly been seriously touched upon.

In general, we increasingly value the contribution of the layperson, in a sense the "ordinary people," for our cultural efforts. The mere technical expert is losing his godlike respect and ungrounded influence, in the face of efforts by young and old to free their minds from dry book learning and expand their souls. *Progress without the work of the layperson is hardly conceivable any more.* We want and need the most careful division of labor and the highest level of specialization. But in addition to solid, indispensable expert knowledge, we also need to instill a sense of recognition or feeling for the worth of *the best* that our life surroundings have to offer. If our tedious professional education doesn't allow us to conclude that the cabinet that we're buying is made of poor wood and with sloppy workmanship, then in a similar manner, in our daily lives we lose in a hundred places what we have gained in one. If a bank director in future doesn't have the least idea of the garden he wants to have laid out—then he is nothing more than a wholly average bank director.

The layperson can already *indirectly* assist with the garden. *Be discriminating!* Don't allow shoddy work from just anyone, especially not with your new project. Demand *quality work*! Reject the jumble of pictures and unprofessional drawings, this disease among our applied arts, that is quickly cobbled together expressly to infatuate you. *Ask to see the thing!* That means for your garden, exact plans, sections, photographs, exact plant lists and cost estimates, and if possible, a bird's-eye view of your future garden. These are all things that you can review and that you need to *believe in,* and not just to be good-natured. And finally—this can't be repeated often enough—*hire your garden designer in the beginning,* when possible at the same time as the architect. Then you will avoid the loss of much of the material and some of the ideal value of your garden, that you would never have known possible!

These are simple, useful statements, which alone will produce results when followed.

Of course, even more valuable is the *direct* work of laypeople in the garden, in their own garden. It is wholly appropriate that it was the artists who created the first beautiful gardens in recent times, but it is also correct that the true garden of our era will first appear once the layperson has made a practical contribution to it. Yes, one may say: *the level and intensity of our garden culture is especially dependent upon how well amateur gardening fulfills the tasks at hand.*

These tasks are diverse. Here we have the broad area of *flower cultivation.* In this case, as good as nothing has taken place on the side of laypeople. The discrete reproofs of Lichtwark, the fine example of Gertrude Jekyll in England, these and other examples only influence an all too limited group. Today, the gardener assumes responsibility for the selection, planting, and tending of flowers. Here he rules as a dictator. But he grows and variegates a flower only for the market, as an expert he hardly recognizes their dormant, deeper qualities. By contrast, the artistically educated amateur more easily perceives the latent color and formal possibilities of a plant because he senses its spiritual expression. He is by far the more objective appraiser, and not least, he also has the leisure to do this.

Improvements in gardening, such as the useless so-called serpentine spruce, tasteless hybridization like white-variegated-leafed cherries or the yellow mountain ash, atrocities like certain purple fuchsias, violet carnations, or the monstrous *foxglove* (Digitalis monstrosa), would find no respect in the eye of the educated amateur gardener. To say nothing of the over-sizing of individual plant organs or the generally superfluous serrations of leaves, and similar madness of the cultivator. On the other hand, in the future of gardening, splendid plant forms such as the *field rose*, the *juniper*, the *pyramidal poplar*, the *elderberry*, the *sunflower*, and others will be recognized or newly discovered by laypeople. All of these latest gardening absurdities from the field of plant cultivation will give way to a rational method of selection, once the laity become more actively involved.

There is one more thing. Such interests, raised to the level of noble passion, are often capable of influencing wider fields than previously thought possible, due to their gentle origin. They have an effect on the economy and on popular education, and strengthen piety. The famous *bulb culture of old Holland*, this nation that still today is the world's supplier, is preferably lay work. We obtain the most beautiful kinds of *irises* and *lilies* from the Japanese, those born plant lovers. The perennial *phlox, poppy, peonies, hollyhocks, asters,* and *anemones*, the summer-blooming *reseda, stocks, wallflowers,* and *rosemary,* which are largely ancient herbal and medicinal plants, have come to us via medieval priestly and cloister estates, and subsequently through the laity. And today England owes its unchallengeable gardening fame not least to the interests of amateurs, as they demonstrate through a formal cult of *orchids, roses, sweet peas, hollyhocks, nasturtiums, bellflowers,* and others, as well as through the staging of large-scale, elegant *horticultural competitions* (Templeshow).[32] In order to achieve such practical results at home, we must also strive for passionate amateur activity in the garden. And in this way, and this is particularly dear to my heart, women will be able to renew their own original field of activity, as in earlier times.

Yes, our women! I am thinking of another kind of "garden work" that may be specially reserved for the woman. This is the *social reconquest of the garden*. Just how

plate 11

An oak allée planted underneath with azaleas: a symphony in gray-green and orange, enriched by the blue and white of the lakeside terrace under the arch.

PLATE 11

much the house garden of our grandfathers' time acted as the social gathering point, we know from Goethe. In those days, women and girls customarily passed their time by lightly working in the garden, when not needed in the house. The children tumbled, dug, and planted in their own special plots; although little city sprouts, they grew up in freedom among the green. And as to the menfolk, a thought or form may have come to some fine mind when contemplating his beloved, rare carnations, which we value today as a precious communal asset. The still evenings, just as the morning dew, found the entire family gathered together, and on Sundays or during celebrations, the green spaces resonated with the murmur of the hundred voices of dear relatives. One criticized and held good conversations, one danced, made music, and acted out jolly theater pieces. Yes, so it was: young and old preferred to go into the garden when they wanted to be *joyful*. Now that is no longer: the lady of the house receives guests in the salon, for goodness' sake, certainly not in the garden, and our loving couples "find" each other in the cinema.

It will not remain this way, for certain, and signs of improvement are already at hand. But that is not enough. In the future, our valued women must promote the garden as much more than just *a domestic facility*. Through her, it must again become a second, open-air home for today's family. Once secured and established, this convivial moment contains within itself something further. Then just as each applied art, house building, or interior design receives its ultimate expression from the kind and intensity of use, so now will newly developing customs and practices in the garden determine its future character. It is a matter of how often within our material existence a specific area of work for the woman may be inserted into the fine weave of artistic and stylistic developments. The poverty of our attempts at new garden design first fully reveal themselves through their still pure *fundamental will* [*Willensgrundlage*]. Here and there, our gardens are outwardly dignified and beautiful, but they still lack harmony among the plants, a more intimate air, and overall any kind of spiritual content. In order to improve this situation, we can do no better than to enlist the aid of noble women. Thus here, in the *conscious spiritualization* of the new garden lies a natural endeavor for the woman, as yet unfulfilled.

In the end, aside from such sociable amateur activity, the garden work of the woman can be thought of as *immediate*, pure horticultural work. One searches so often today for positive and promising professions for female youth. Generally one thinks of beautiful nursery gardens last in this connection. And even then more in the sense of a "vacation camp" or under the motto: "something different." The horticultural business now already has a whole range of positions that intelligent women could fill very well, and to the highest degree of satisfaction. How often when I and my trusty staff are buried over our heads in garden plans, have I wished for female work assistants to lend to us hopelessly formal, rational men, their light touch in composing with flowers and other things, along with their sensitive souls and their God-given serenity. If our women would make themselves more familiar with the promising future of gardening, then there would be far fewer of the "misunderstood," and yet something more—beautiful women.

Thus for our garden culture *how* a person stands in relation to the garden is of decisive importance. If he does as he pleases—from a technical, economic, or rhythmic standpoint—behaves passively or even indifferently, then even the best intentions of the leader won't help. *The individual is absolutely indispensable.* When the agents here in question—the gardener, the architect, and the layperson—come together with a proud and eager spirit, at whatever cost, to achieve gardens worthy of our day: *then everything is possible!* Within the collaborative work of these three agents, which comprises the entire intellectual and material content of the garden, lie hidden its next phases of development.

Given time, with intellect and labor a special garden-psyche may be formulated, with every impulsive, definite feeling for everything concerning the garden, which among the current generation is now primarily found in England. *Modern man, bound by technology and science, in his desire for liberating rhythms and his instinctive brotherly love—he stands up, and makes his own garden!*

V

VEGETATION

A. The Visually Striking among Plants

In the end, the intrinsic value of a garden depends upon the particular manner in which we utilize the plants within it. Not only plants as the means of fulfilling specific garden needs, but even more so plants for their rhythmic effect among one another: color and form *as such* are decisive here. Color and form are judged by our eye as they are reflected in it. The eye of a tradition-laden Western European is different to that of a guild master in the Middle Ages, or of a Berliner in the era of Frederick the Great. And none of these have the naive openness of a freshly baked Canadian, as he looks at the world about him.

We have experienced the new natural sciences: *our eye has grown more subtle,* more calculating, it sees the outward appearance of the plant at the same time as its inner structure.[33] The general evolution of technology has roared over us, with its mass production and unusual constructions: our eye is accustomed to strong contrasts, perceiving *more freely,* it feels the monotonous one-by-one of the trees of the forest, just as greatly, as it delights in the unlimited boundlessness of the sea, or as it rests upon the beautiful waste of the heath or the colorful expanse of the desert—we are new people! As such, we are on the point of ultimately transforming and refining our communal life as well, and we are now sampling the refinements of a comfortable way of dwelling and living: our eye partakes in the *tender harmonies* of a softly lit domestic interior. And our eye, this incredible wonder, in addition to all that, is capable of royally feasting upon the surface of the superfine skin of a peach, or the clear freshness of a dewdrop on a flower petal. What is our eye?

Doubtlessly, as much as the garden will be defined by obvious human needs, especially in the coming future, in any case its more refined and ultimate development will be strongly influenced by: how we *see* things. Our new eye, led by as yet unprocessed perceptions, will decide.

The means to this end in the garden are the plants. It suits this purpose well, that today we have more plant species for our gardens than was ever before the case. Not only numerically, since there are more gardens, but because we possess

PLATE 12

an incomparably greater number of *new* plant species overall. Our plant cultivation has incorporated a large quantity of new species in the past decades, and variegated them a thousand times over. The red maple, for example, the clematis,[vi] most of our beautiful shrubs, fine cherry and apple varieties, still haven't been thoroughly exploited by French and Dutch garden designers, and the splendid lilies, irises and Nymphaea, silver lindens, red oaks, Sierra redwoods, rhododendrons, azaleas, which a vast import system has brought to us from foreign countries (Japan and North America, the Orient), were also previously unknown to them. Science, trade, and technology have worked together to provide us with new, completely different products from beautiful nature. Our gardens are plant-millionaires, and our plants are cosmopolitans.

But this overwhelming abundance of nature's bounty is also the thing that makes our gardens poor, and we stand helpless in the face of it. Therefore, the organization of plant materials in the garden is one of our most important tasks.

Those are a few signs that we must heed. In general, we should accustom ourselves more to the thought that our gardens demand another face just by virtue of their new materials, and we must therefore fulfill our responsibility: *to utilize.* To experiment and take risks with plants and flowers is the task of those who are obliged or inclined to serve the garden of the future. These days we see a multitude of formally successful garden constructions, but only a few of these represent *the self-will of plants,* either in combination with them or wholly independently. Therefore once again, use them without ceasing! The necessary influence of plants will also determine the form of the garden, as soon as we have first begun to use them exhaustively.

When I expand this discussion by speaking of vegetation in particular, then naturally I mean to give preference to the *new* vegetation. New plants or even a new style of using them—in the sense of the self-delimited purpose of these pages: to discover within today's garden forms, the essential old and the potential new in respect to our future garden. Thus to this end, I deliberately avoid following the obligatory list of appropriate plants for this or that purpose. Real gardening knowledge is not attained through calculating machines; the true gardener should work on his own. I am only giving mere suggestions here, as I briefly review the material of the garden through random samples—I know that there is much more besides.

What goes together well there, in consideration of the aforementioned viewpoints?

Let us imagine a gray-green lawn area (one can make this using special grass mixes on particular soils), with white and yellow narcissi springing up within it,

vi This book is not least written for the laity as well. Therefore I give the plant names in Latin or German, according to which seems more common usage to me.

plate 12

Our garden may also be pleasant in the wintertime. Here the architectural garden presents itself particularly well in frost and snow.

enclosed by a thick, high wall of yellow-, brown-, and pink-colored maple species—sprouting maples formally resemble flowering trees. Or, an open grove of dark red thorn stems, under whose elegant silhouette white chrysanthemums stand. And further: a wreath of blue-gray Chamaecyparis squarrosa [Japanese false cypress] at a man's height, encircling a plain of flowers, with a mixture of snowdrops, scillas, colorful crocuses, harmonizing with primroses and auriculas, tulips of yellow, red, and other colors besides, along with pure horned violets, blooming from winter snow thaw to the turning of leaves in autumn; low stems of Buxus spheres accentuate this plain.

These are garden images, formed by plants alone, colorful garden harmonies. In this sense, birches go well with blooming, light pink climbing wild roses, and orange-colored azaleas are lovely with laburnum, and these with purple wisteria, or with white wisteria climbing over pink-flowering locusts. How tenderly the forget-me-nots sit under the apple blossoms, and the white lilies among soft blue bellflowers. And a bed of golden brown late iris bulbs with baby's breath trembling above it, is a small floristic feast in itself.

But even more striking, and often more welcome outdoors than this soft symphony of colors, are the beautiful *contrasts of color*, bestowed upon us by plants. There we have purple and blue autumn asters in front of a glowing wall of Boston ivy, or coupled with rust red Japanese larches. On a white carpet of Arabis, dark red tulips. A large circular bed is also imaginable, bordered by white, yellow, cream-colored, carmine, and red roses in increasingly narrow rings, progressively rising to a small group of white stems: an enormous, ornamental bouquet, for which the selection of varieties is almost incidental. And in order to get beyond flowers, what about a grove of flaming yellow plane trees surrounded by an Ilex hedge?

These are examples that I have deliberately chosen in order to show the endless variety of the entire field, which includes the traditional. Thus as everyone knows, roses go well with ivy, and flower borders adjoining a path are a frequently used garden motif.

The *form* of plants can even more characteristically determine the garden image than the colors. Moreover, this is the surest effect: nature is not really dependent upon colors. To begin with, I will consider only the many forms of nature.

A large oval of dark pyramidal English oaks steeply crowns an embankment, while over this terraced bluff, forsythias, boxthorn, clematis, and nasturtiums softly fall like waves of flowers; by the edge of the lower terrace, this rhythm of forms is reflected in the mirror of a pond. Or a path is bordered by hornbeam, out of which grow hawthorn spheres. Yes, each allée and every pergola, in the end, is nothing other than a garden form brought about by plants *alone*.

Indeed, it is also conceivable that the planting themes of the architectural garden composition may be wholly or partly derived from nature's precedents. Blue liverwort is known to prefer growing where white anemones bloom under gray beeches; the golden blooms of gorse glow particularly splendidly by the gray-green of junipers, and spruce glades, sunflowers, and foxgloves represent a rhythmic unity, which apparently cannot be imitated by other means. And so there are many beautiful, even refined form and color compositions outdoors; nature is never tasteless. But it is also not always interesting, as we know. It is this inner stimulation, this delightful awakening of our perceptions and nerves, that we expect from the more intensive imagery of the garden in contrast to nature beyond it. Here is the human application of rhythmically intensified, *translated* nature. These examples are only natural through their inner harmony. It is therefore mistaken to think that because of the self-evident appropriateness of nature to garden planting concepts, it must be necessary to want to imitate it in some form. However, more on this later.

But by far the most striking contrasts of form and color in the garden are achieved through more functional compositions, or where *building structures and vegetation* are interwoven in some way, a method whereby either the form or color of the plants, or their structure, can alternatively dominate. The prior example occurs when we position a basin with an interior edged with blue water forget-me-nots within a square of delicate weeping golden willows between silver-gray poplars. If one will, this is the elevation of an architectural garden ornament to a garden concept. The same occurs when I enclose a play area with a thick hedge; it immediately becomes a living room. Further, a narrow strip of water surrounded by lawn, with arches covered with red and white roses standing over it, primarily owes its beauty to plants.

More "greened architecture" comes about when I make this canal much wider, planting it with water lilies, covering its gray walls in red vine leaves, accompanying the whole scene with pyramidal poplars, and concluding it with a cascade. Less splendid but no less impressive, are yellow tea roses trained over the green trellis-work of a pergola. And the idea of a lawn tennis court that I enclose with a high wall of arborvitae, is very pleasant indeed. Each pergola, every espalier fruit wall and house wall, is an example of the potential marriage of architecture and nature in the garden, for the possibilities in this direction are legion.

Clipped plants are also garden architecture in a figurative sense, which we use to express man's freedom, allowing them to grow to a prescribed form.

As much as we feel the rhythmic effect of this kind of bounded nature in its appropriate place, we also openly recognize the characteristic beauty of freely growing plants. Yes, it is without a doubt our garden-architectonically trained eye in contrast to the former "natural gardening perspective" that—as every professional knows—presented and maintained its material as unnaturally as possible; it is *our eye cultivated in respect to nature*, that appreciates the worth and finest qualities of freely growing plants. The real, irreplaceable beauty of the laburnum, the wild rose, or the blackberry shrub, the ornamental expression of chestnut and

PLATE 13

tulip trees, the juniper, the Taxus, and the stock rose, the fine-lined silhouette and charming habits of birches, locusts, and swamp cypresses; once we have more or less "seen" these, then we may easily consider them to be our own discovery. And with the literal digging up of such "ordinary" Cinderellas of the landscape period, such as the elderberry, the sumac, the poplar, nasturtiums, foxgloves, and many other plants of truly spiritual expressive power, here I would really like to see evidence of our era's ability to pursue the highest form of garden culture.

This is how an eye considered normal these days sees things in the garden, how varied our material is, and how rich the possibilities of utilizing it are. There is scarcely a product of a vivid imagination that cannot be realized through today's garden materials and potentials. Much more, the contemporary garden needs a great number of vegetal fantasies; in respect to vegetal concepts it has doubtlessly been neglected.

But I have something else to say in this excursion through the workshops of the garden: the main objection of serious opponents of the architectural garden is always that it is too *one-sided* in terms of vegetation, and that the obvious inbreeding will lead to its demise in the end. But all of the vegetal motifs that I have introduced, as natural as they are, were nevertheless consistently *rational* structures. This endless abundance of vegetal capacities, this unaccustomed, overwhelming wealth, demands rigorous representation through a controlled composition. This is only fully realized through *the spatially conceived garden*, which will certainly not suffer from boredom.

Precision and wise restraint are the qualities of this garden form by which it has spread everywhere of itself, and which will also have an influence on the utilization of garden materials. Today, the garden that displays the greatest variety of planting is often considered the most beautiful. The time must come when the beauty of the garden is valued according to the strength of *all that is living* within it. Compose and plant to no end—but be disciplined!

B. Applications

TREES AND SHRUBS What particularly distinguishes the use of these in contemporary gardens in contrast to the appearance of earlier ones is the newly emerging reaction against the capricious, gaudy mixing of species in masses and clumps, as was common practice in landscape gardens. We now cultivate more of an intelligent *sense of limitation* in this respect, which at the same time guarantees higher pleasure. We achieve this in one and the same garden, by perhaps only

plate 13

A truly successful front garden, which, despite its open situation on the street, still can be used for socializing.

using spruce and maple species as focal points, while in another, only oaks, locusts and ornamental cherry trees provide the necessary shade and shelter. Through such limited selections that naturally must also incorporate smaller vegetation, we not only intensify the expressions of the plants themselves, we thus at the same create a necessary, quiet background for the varied forms of the singular specimens within the garden composition. The collective plant materials of the garden should thus as far as possible be in *essential related harmony*, for example, *mountain ashes* with *birches* and autumn shrubbery, or even in *harmonious contrast*, for example, the overhanging, fresh green, pinnate leaves of the *tree of heaven* and the *wingnut* against the solid *hornbeam* and *linden* or dark *Taxus*. In the same way with color: such as the golden yellow *poplar* with the bright green *beech*, or the *ash-leafed maple* against *alders* or *spruces*. Our garden may derive its special character from such thoughtful combinations of large structure-forming materials from the outset, which in spite of its external uniformity still distinguish it from a hundred others. Under the chaotic hodgepodge that ruled previously, one garden was just like the next in respect to vegetation.

Following these principles, we approach the composition of shrubs and the so-called tree-like varieties with even more rigor where possible. The recent appearance of lasting perennials has dealt a deathblow to the traditional "bosquet." Here from the beginning, we will exclude a series of less attractive blooming species that were forced upon us by a century of overzealous breeding. The use of Loniceras, weigelias, caraganas, snowberries, etc., is to be limited in the typical garden of today, not to mention the even less common buckthorns, amorpha [false indigos], bladder sennas, and others. Also most of the varieties of the otherwise useful *deutzias, mock oranges, ribes, privets, hydrangeas, forsythias, cornus, berberis,* and many others, cannot be well represented in our gardens in their full number. It is truly monstrous how here, as well as with some tree species, how an indiscriminate hybridization raged among us, ruining otherwise pleasant plant forms. In the catalogue of one of our leading nurseries, I count, for example, about fifty summer-blooming spirea varieties, and then one thinks: fifty different summer spirea, one like the other—a wholly uninteresting twig-broom!

In contrast to this official mis-breeding of shrubs (older dendrologists already list more than ten thousand species of conifers), we again turn to *beautiful selection*. Then the *lilac, laburnum, hawthorn, ornamental apple, and cherry varieties are* indeed beauties. Here we may variegate wonderful rhododendrons and azaleas to our heart's content, but again with intelligent limitation. We accept the true outsiders, like the *mahonia* or the *elderberry* (but the "common" Sambucus nigra), and we raise no objection when considering the hundreds of *wild rose* varieties: indeed, even when limited in this way, within this realm of woody plants there are still new and valuable discoveries to be made for the garden. But we don't just haphazardly stick these true, proper *blooming* shrubs together and on top of each other, instead we use them to form interwoven groups and open hedges. And we place the highest values on the *individual shrub* just as we do with the *individual tree,* in respect to its beautiful free growth, the purity of its colors, and the unrestrained power of its unique appearance.

HEDGES The trimming of plants into regular forms has also found many opponents within our movement. But we need only observe the windbreak hedge on the shore or the well-known farmhouse lindens, in order to recognize that there are influences that naturally call for the pruning of plants into fixed geometric outlines, and we need not turn to enclosing hedges only as protection against dust and wind, or privacy alone. Thus, for example, I always came to the same conclusion with the garden plan of a cooperative garden city with small spaces: hedges. Simply because there was no space there to sensibly include large tree forms, except occasionally. And if trimmed plants themselves make so many practical things possible, then logically we need not shy away from the conscious use of trimmed plants for *higher purposes,* in a certain sense, sometimes even for purely rhythmic reasons. When this exercise is intelligently limited to the elementary geometric forms of rectangular and right-angled prisms, the sphere or the pyramid, without degenerating into gimmickry, then we scarcely have anything to fear for our garden pleasure. To the contrary, through shaped trees and shrubs we often possess a means of limiting stone and wood elements in favor of the materials natural to the garden.

Good hedges are formed by the common *privet* (Lygustrum vulgare), as well as the oval-leafed (Ligustrum ovalifolium), and the red-berried cotoneaster (Cotoneaster simonsii) also makes attractive hedges, with red berries that in autumn glow like coral among their dark green foliage. The *hawthorn* and the *blackthorn* are ancient, often used hedge plants. For lower hedge forms, most of the short-shooted shrubs like *ribes, cornus, berberis, mock orange* varieties, and others may indeed be used if necessary. Yes, with the proper treatment, even two- to three-year-old oaks can form attractive ornamental hedges. But the most perfect and versatile hedges are of *hornbeam,* which in suitable climates have been used for such purposes for ages. There is scarcely any kind of hedge from the bedding enclosure to the high wall, which this extremely eager and highly useful plant cannot form. More sensitive, if also especially beautiful, is the *beech,* which fully develops only under the best circumstances; when placed in dense urban areas and in windy conditions, for example, it is always sickly. Among the larger and largest hedge forms comes the *linden,* which one may use to advantage in its large-leafed form (Tilia platiphyllos), and which can be grown to the height of a house. Even better suited in this case is the long-shooted, bright green Crimean Linden; the small-leafed (Tilia parvifolia) particularly for low hedges and pergolas.

We will not discuss the cultivation of hedges here. But it must be mentioned, that once planted (in well-prepared soil!) they need to be well maintained, trimmed, and fertilized for years in order to grow quickly together and be dense all the way to the base when mature. Without maintenance, even the so-called instant hedge shrubs will not make hardy hedges.

Another category in this brief review of hedges are *conifers*, with dense natural forms inherently suitable for hedges. The old tradition of the *boxwood* hedge, this screening plant par excellence, is well known. The standard *arborvitae* (Thuja occidentalis) is also particularly suitable for hedges, as is the *yew* (Taxus baccata). To my knowledge, the latter belongs among the most resistant plants overall; dust, soot, and shadows have relatively little effect, it is only sensitive to wind, too much rain, and to drought, like all conifers. And in the case of dreaded shadows, other plants such as ivy and asarum thrive there, as well as for example, ribes, hawthorn, hornbeam, and even privet, when they are grown *with* the shadows. A hawthorn hedge planted under lindens at the same time as them, sprouts just as freshly after a lifetime as in its first year, when its roots are protected from those of the trees by waterproof roofing paper, and it is kept well fertilized and watered.

One kind of hedge that has not been given enough attention is the so-called *free-form hedge* composed of shrubs that although densely planted, are either lightly clipped, or not at all—a good replacement for ugly clumps of shrubs. Among these, the *mahonia* makes lower evergreen forms, and the Ilex higher ones. Broad-leafed woody plants are even more attractive for this purpose. For the so-called *blooming hedge* my first choice is the *wild rose*, among which the sweetbriar is one of the few that also blooms when clipped. The various spring bloomers like *spireas, Japanese quince, forsythias, deutzias,* and the evergreen *rhododendron,* are also particularly well suited here. In planting such blooming hedge plants along both sides of a path, we create *hedge corridors,* for which we may use taller species like *lilac, elderberry,* and *laburnum.*

Finally, the dreaded "stiff hedges" are not always so bad. The nature lover can find sympathy for their strictest forms, when he realizes that many of our best *songbirds* prefer to nest in the impenetrable dense twigs within these hedges. More recently, we have begun to plant so-called bird protection hedges, where our larger or higher-nesting feathered singers happily have already found a suitable home. Therefore: plant hedges!

ARBOR WALKWAYS These shady green ambulatories are actually nothing more than overgrown hedges, so to speak, trimmed in particular forms. Although there are enough examples where outsiders like hawthorns, chestnuts, beeches, and others may have been successfully forced to grow in arbor-like forms, *lindens* and *hornbeam* are the given materials for pergolas. Therefore we usually use hornbeam for closely branched and low forms, and the linden for larger forms. Both adapt themselves willingly to the most varied forms, which have gradually evolved in response to differing needs for shade and protection. The simplest among these is the arbor walkway that is closed on the sides and above, with a semicircular or rectangular section. If we need more light, especially for longer installations, then we can open one or both sidewalls with windows or slits. When an airy quality is needed, then I like to plant young, uncut, multistemmed linden saplings (*Heister*)[34] spaced singly or in pairs, or three together. This then becomes nothing less than a green protective roof carried by pillars, that looks as precious as an arcaded walk in an old cloister.

ALLÉES When this kind of freedom is taken somewhat further, then we have the allée. We all know what an allée looks like and also that it can be made with almost any kind of tree. I only mention these here in order to show how this beautiful form became degenerated during the landscape period. Even today, the real garden expert plants his allée wide enough, so that, as he says: "each tree can still develop its full beauty." This is but a complete misunderstanding of the term "allée." *Within these it is much more the case of the single tree giving up its individuality to support a complete, new organism.* Seen from its outside, a true allée presents a closed green wall, and from the inside a more or less open, high or wide, cathedral-like vault. Thus, for this purpose, the trees must be planted considerably closer together than required for their natural development; they must grow upward in unison. The general correct rule here is that a closer spacing between the trees intensifies the speed of their development, and often the uniqueness of the form as well. Following these requirements, one can plant poplars and chestnuts relatively far apart, i.e., not more than eight to ten meters, and elms and maples, five to six meters. Indeed, I have found *linden allées* as high as a house enclosing a moderately wide path, with trees not more than two to three meters apart. The view was outstandingly beautiful. With similar spacing, one can unify our distinctive *spherical forms* of *elms, maples, locusts,* and others, into allées with crowns that form a wavy line from the outside, and a green ceiling on the inside. Obviously these sharply outlined spheres, as well as the *pyramids* of *poplars, oaks, and elms* with their yellow and silver-gray varieties, present independent and proper rhythms, even when planted farther apart. But still only as pyramids and spheres. With these tree forms and also with birches, under certain circumstances one can on the other hand go down to the smallest spacing of one to two one to two meters, in order to attain original effects.

As an example of how allées should *not* be planted, that is as a counterexample, I choose from out of a hundred others the lake promenade [*Teichpartie*] in front of the restaurant of the Bremen People's Park. There, four-row deep allées are found, with the two inner rows of lindens and the two outer rows of chestnuts, the whole surrounding a lake. That this difference in materials is already a disadvantage is made even clearer when one measures the distance between the trees. The lindens stand six meters apart and the chestnuts—as larger-crowned trees in the best landscape garden style—at about twelve meters. Together this results in an unevenness and a dissatisfaction from every imaginable standpoint, that is only surpassed by the completely inadequate sense of enclosure created by the lindens by the lake, with crowns pruned as high as trees along a highway, crowding out the thorn pyramids unnecessarily planted under them. In addition, the whole arrangement is a striking example of the fact that an architectonic layout without an architectonic *spirit,* defeats the purpose, place, and sense.

GROVES If we place a number of allées together in a row, then we create a grove. It hasn't really been accepted yet, not even within the newer garden creations. Outside of historical examples, we actually only know of it from the tavern garden. And yet it is probably the one garden form with which we can make the deepest impressions. The accentuated tree-on-tree of the grove, the flickering play of shadows on the ground under it, the plastic clarity of its external appearance, are things that have made the grove a preferred and typical garden element of important epochs for ages. The antique world and the orient laid out groves in which they honored their greatest, and even today the Japanese and Chinese dedicate holy groves to their gods. And aside from this symbolism, could there be a more beautiful garden concept than that of observing the deep shade of a *grove of plane trees* (Darmstadt), or the solemn semidarkness of a grove of *birches* from the sunny cheerfulness of flower oases and the small waters of our gardens?

VINES The vines in a garden comprise a chapter on their own, which couldn't be more different from the preceding. If the dense grove and the pergola need substantial time to create a scene, then the swift growth of vines is ready all the more quickly. There is no similar means of evoking a strong impression in the garden within such a short time other than vines. Thus, their potential applications are so numerous that if I so desired, I could make an entire garden with vine-work and not a bad one at that: beginning with the house. I would cover its south and east walls with *wisteria* and yellow *roses,* and the west and north sides with *honeysuckle* and *ivy.* A projecting veranda could support *clematis vines* and an open-covered terrace looking toward the west, *wild grapevines.* Beginning with the terrace, an elevated pergola over the main path of the garden would enclose one side; it would be covered with the graceful vines of the common wine grape (Vitis vinifera), and across from it, *wild clematis* and *aristolochia* would climb along a light espalier on the other boundary. At the point where the lawn leading from the house gives way to the kitchen garden, the enclosures on either side connect with vine trelliswork covering the paths and forming large squares, here using white and dark red roses. Between the trunks of a row of trees at the back, the five-leaf Akebia

plate 14

An arbor walk (vine-covered frame) open above and to the sides, accompanied by a flower border.

PLATE 14

forms a curtain. *American bittersweet* and *actinidia* climb here and in groves further back. On a retaining wall that separates the utility garden from a bluff, *Boston ivy* and *Vitis muralis* cling to the retaining wall that separates an overhang from the kitchen garden, both with red foliage in autumn. Here, and in the garden as small pyramids, columns, and open arches, we also find the varied and diverse, beautiful summer-blooming vines, *sweet peas, nasturtiums, cobaea, Maurandia, scarlet runner beans,* etc., while *blackberries,* long-shooted *boxthorn,* and *wild roses* proliferate on the enclosing walls of the orchard. One sees that there need not be a lack of either flowers or shade in our fantastic, quickly growing garden of creepers, which was only described in more detail in order to properly emphasize the value of this material for the garden.

FLOWERS The light character of creepers leads of itself to the even more charming world of *flowers*. What would our cold Nordic gardens be without sparkling floral ornaments! It is their reason for being and their very substance. I believe that the first impulse toward our new garden movement resulted from the unpleasant realization that the former way of gardening had begun to completely ignore flowers. There was enough of pristine green lawns and shrubbery, more than enough. But in our case, we want flowers.

One of the oldest means of bringing flowers into the garden is *the border*. Within this garden form we indeed can show to best advantage annual summer flowers and bedding flowers, as well as roses, but most often we think of the multitude of lasting *perennials* when we speak of flower borders. In recent times, it was England that brought this garden element back to an honorable position, as well as to a certain level of culture. One says that the English example shows a particularly fortunate overall sense of tact for color combinations. And that is in fact the case. However, during my studies on location, only in a few places was I able to detect an obvious and consciously conceived color harmony within flower compositions. The generally good and never unpleasant effect of their borders is attained by the English much more through a rich mixture of species from the beginning, interspersed with more medicinal and herbal species with neutral flowers like *lavatera, sage, achillea, veronica, hypericum,* and others, and not least through constant supplements from the greenhouse during the summer. Although a better orientation and level of experience are not to be denied, nevertheless, it is for the most part only wise prudence that assists the English garden enthusiast with his much-admired flower-lined walks.

We, who today still work on average with much more limited means, while still approaching garden problems more freshly and with more daring, we must follow other paths as to questions of detail. I have found through years of practical experience that a certain kind of *interrupted symmetry* in their planting improves and often enriches such borders. With this more rigorous organization we not only have the forms more in hand, but even more so the color expression. In regard to colors, I like to choose strong tones, placing white in between for order and contrast. For the true mixed border, it is advantageous to group species according to their *flowering period,* if again only as a general rule. Their *height* is also to be observed; for practical reasons and to create stronger shadow patterns, I love to place the larger growing, higher perennials next to lower species, as well as next to widely spaced, close growing or autumn-blooming species such as *lilies, tritomas,* and *autumn asters,* underplanted with low summer and spring flowers like *Papaver nudicaule* [*Icelandic poppy*], horned violet (Viola cornuta), *primroses, daisies, Arabis,* or *carnations.* Very often I further heighten this contrast in form and color by interspersing evergreen *Buxus, Taxus,* and *Ilex.* Wherever possible, a variety of nice summer flowers should be used to fill in the gaps that are always left by spring bloomers in such perennial borders. Thus treated, our perennial border can ornament the garden throughout the year with a constantly changing appearance, and still be completely different from those of the neighbors. (See plan below.)

I usually refer to a particular variation on this form of flower composition as the "evergreen border." This is appropriate for the semi-shaded and fully shaded areas of the garden. The whole series of shade-tolerant perennials, such as *primroses,*

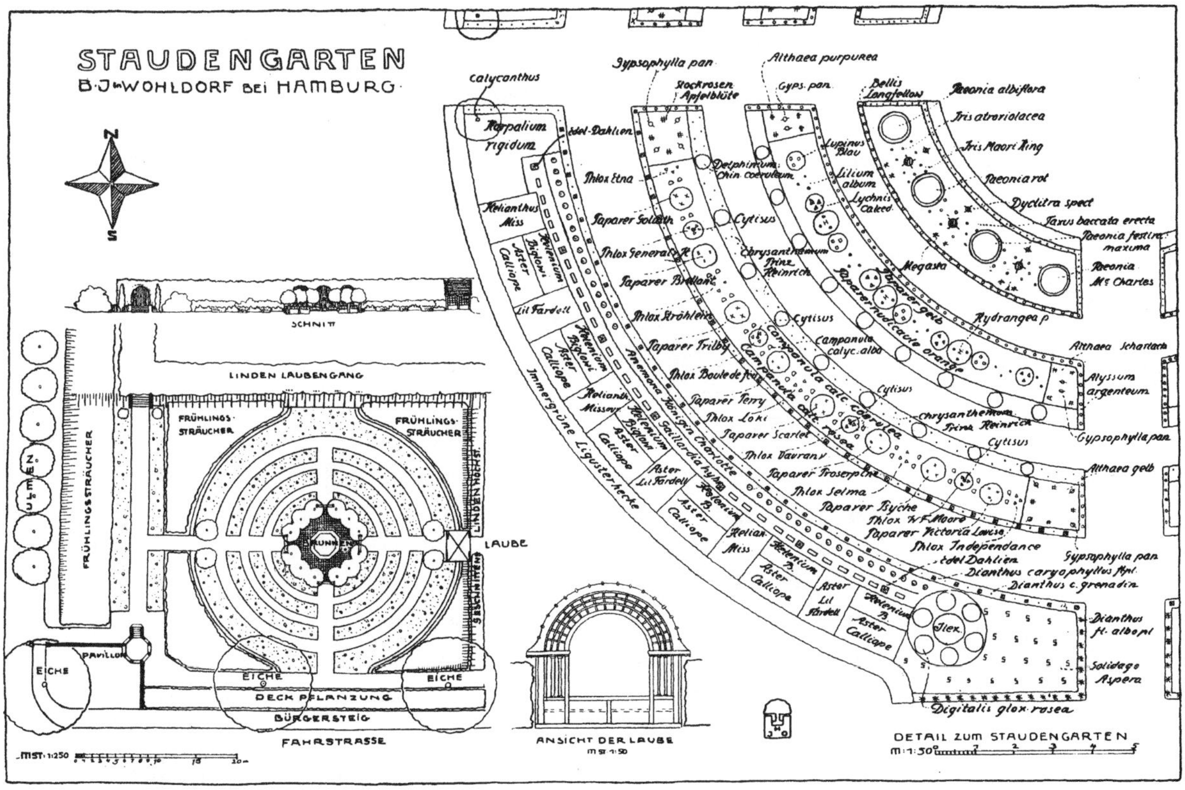

Perennial garden in Wohldorf near Hamburg
L. Migge

violets, liverworts, anemones, foxgloves, and others may be used here. However, in this form of border *rhododendrons, azaleas, Ilex, Laurus, mahonia, cotoneaster,* and other evergreen and other semi-evergreen woody shrubs should dominate. Also smaller conifers like *Canadian hemlocks* (Tsuga canadensis), *dwarf Norway spruces, mugo pines,* and many others belong here, although these like other species must be handled with refined discrimination on site. Particularly suitable ornamentals for this kind of composition, in which I also use *ivy, winter jasmine,* and other vines, are the varied spring shrubs, such as *cherry* and *apple varieties* (also half-standard), *Japanese quince, forsythia, spring spireas,* and for sunnier locations, *laburnum, hawthorn,* and *lilac* as well. Finally, *bulb plants* are not to be forgotten, for they thrive better here than in normal perennial borders, where they are very difficult to manage. (See plan p. 152.)

FLOWER GARDENS If we join together a number of different beds just of this type, or along with rose beds, thus forming squares, rectangles, circles, or sections of these in combination, then we have a flower garden. Here from the beginning we don't want to be without a full selection of summer flowers. In respect to their readiness to bloom, and unusual colors or forms, such varieties as *dimorphotheca, snapdragons, impatiens, amaranth, nasturtiums,* and *sunflowers* don't find any real competition among the outstanding perennials. It is necessary from the initial siting of the flower garden to ensure a *sunny* location, and that the flowers have a dense *green background* such as a green wall or hedge providing protection and visual relief. A *sunken* location is also useful for creating a good impression for it allows part or all of the garden to be overlooked from a higher position. The composition will develop excellently following these points, whether we want the garden to be for *cut flowers* or to be a *beautiful showpiece.* In the former case, we may observe purely functional requirements when organizing the flowers, sometimes almost without giving it any thought. In general, however, one cannot expect rewarding gardens if the flowers are arbitrarily arranged, even if we cannot quite bring ourselves to call them ugly.

As an aside, it may be noted above all that plants in any configuration never really detract from the peaceful, deep consolation that the poor in spirit find in the garden. Generations of garden visitors have faithfully relied upon the never failing compassion of nature for this reason.

Strong effects in flower gardens arise from *conscious* gradations of form and color, at best when following a guiding idea. Some themes derive from *form,* such as when we surround a central fountain with concentric beds punctuated by archways along the radius. In the case of flowers, naturally the will toward the disciplined use of color is even more important. These kinds of *color gardens* in their strictest, i.e., their actual sense, have scarcely been accepted among us. In fact, the imposed requirement of planting only a few species that generally have a limited blooming period, presumes a mastery that is not really easy to exercise, particularly in the garden. While this garden form makes reference to the greater movements from which it derives, it also allows for the more ordinary kind of flower enjoyment in addition. Thus conceived, a breathtaking sight results when we compose a "blue garden" of *larkspur, bellflowers, snapdragons, petunias, heliotrope, stocks, lobelia, horned violets,* along with vines such as *clematis, morning glory,* and *periwinkles.* Even so, with such experiments one should neither proceed indiscriminately,—such as by planting these species, including China asters or cineraria, in fully foreign habitats—nor pedantically on the other hand. Thus, within this blue garden, I would even intersperse mediating grey-leafed plants, such as *stachys, sedum, centaurea,* and also higher shrubs such as *silver poplar, sea buckthorn,* and others, if not even white flowers as well. Such an unusual creation as the color garden must above all bear the mark of good taste.

Also, a "yellow garden" would not only be made using nasturtiums, sunflowers, and yellow poppies, for we would also include yellow *pyrethrum, eschscholzia, marigolds, zinnias, scabiosa,* and *portulaca,* with *coleus,* yellow *dogwood,* and *elderberry,* running through the entire scale from softest cream to deepest orange. Directly under the scepter of the color yellow, represented more than any other among our flowers, it is also easy to lay out a flower garden that is dominated by this tone throughout the entire year, when we don't always count on a uniform effect. For this, we need only add yellow *crocuses, alyssum, wallflowers, pansies, goldenrods, gold-balls,* and *dahlias.*

Color gardens are also created in a more limited sense, when we restrict our garden to certain well-formed, highly branched species. A yellow or yellow-brown-blue *primrose* garden in spring is a wonderfully beautiful thing, and the glowing, colorful splendor of the richly varied red *phlox* won't be surpassed by anything in summer. An *autumn aster garden* of blue-violet, with accompanying white and pink undertones, offers enough variety to keep our interest from ever being diminished. One can also create beautiful little gardens with the numerous species of *iris, lily,* and oriental *poppy*. Of course one will carefully introduce these theme gardens within the overall garden composition, so that flowers in bloom are not disturbing in any way when they appear.

However, as beautiful and varied as these combinations of flowers and color in our garden may be, there is another material that where possible, evokes an even finer and more impressive effect. That is, *wildflowers*. Previously, one saw wildflowers only occasionally and timidly used in landscape parks or bosquets, or at the end of deep perspectives. That these may appear to particularly good effect within the architectonic outlines of our spatially conceived larger gardens, should be considered here. Let's picture in our minds an extensive park. Following the foreground formed by lawns, water areas, and gardens near the house we imagine that as through the frame of a cross-allée, we see a softly rising slope with

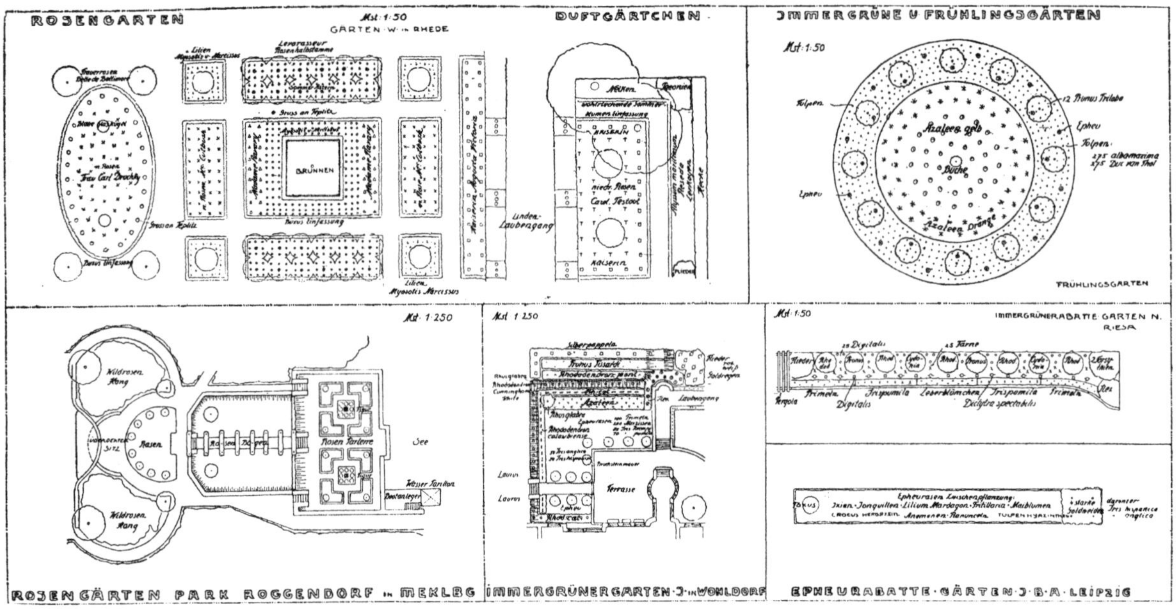

Evergreen, scent, and rose gardens
L. Migge

flowers, a powerful harmony of color, in yellow, red, or blue, or more of these strong colors, all mixed together at once. Farther on, this sea of flowers flows into an area of large-scale beds and agricultural fields, sewn with carefully alternating thickly sown *lupines, crimson flax, blue phacelia, lychnis, sainfoin* varieties, and various other beautiful flowers typical of the agricultural landscape, each covered in one variety. In addition, many summer flower varieties such as *carnations, tickseeds, cornflowers, godetia,* and especially the many annual *poppy varieties* may be used in this wild garden, which could certainly be added to the nursery or cut flower garden to practically serve this effect. By the same method, we could also transform the fields that we include within the open areas of this park into actual *flower meadows.*

Admittedly these are only verbal examples; but we may form an idea of the dazzling impression that orderly masses of flowers in the garden may bring about when we visit the commercial flower fields of Erfurt or Quedlinburg, or the tulip and hyacinth fields of Holland in spring. It happens repeatedly: we are too cautious and conservative to take advantage of the wonderful freedom that today's flower selection offers us, which we should confidently exploit to the fullest, in every direction. But this also shows us, namely, that a great Maecenas[35] of the garden still doesn't exist among us.

plate 15

The simplest flower gardens are the best.

PLATE 15

Only one theme garden remains to be discussed, that on a superficial level may seem to be the most opposed to any kind of mass development—because it is the least visible. Nevertheless to my mind, the little *scent garden* represents one of the finest and most pleasurable stimulations in garden living. (See plan p. 152.) *Thyme, lavender, mint, basil,* and *rosemary* (in pots) are ancient spice and scent herbs. But also many of the so-called spring bloomers, like *wallflowers, hyacinths, mayflowers,* or summer flowers like *reseda, stocks,* and *sweet peas,* are appropriate for this purpose, as are many lilies. Those who have a strong nose or passionate interest in this direction, can also tolerate *tobacco, marigolds,* or *tuberoses.* Among shrubs, *daphne, lilac, oleaster,* and the *spice bush,* have strong scents. There is no reason why we shouldn't even furnish our scent gardens with *lindens* and *bird cherries.* But we would seldom make such a great effort here. It is often enough, just to incorporate modest herbs and small flowers within borders lying against the wind near the house, that will fill the evening air with their fragrances.

Also the **ROSE GARDENS** (see plan p. 152) which we encounter the most often and therefore mention the last, give off a vain perfume. Or indeed they should. It well appears that in recent times scent has been given far less attention than pleasing forms or colors. But beautifully scented roses are still found, especially among the older breeds. I may name: *Étoile de Lyon, Frau Geheimrat von Boch, Herzogin Maria von Ratibor, Madame Jean Dupuy, Mme. Jules Gravereaux, Gloire de Dijon,* and others, as well as the truly old *Maréchal Niel, Niphetos, Souvenir de la Malmaison,* and many more. One should select these, not least, in order to persuade the breeder to pay more attention than before to the *scent* of the rose as one of its most valuable qualities. For in their form and color, in their overall essence, I value *lilies, bellflowers, hollyhocks,* or a *blooming apple branch* almost more than the noble rose, and it would be the same for me in other cases. But nothing among the great creations of nature is so rare, nothing approaches the pleasure of knowing that one single plant species is able to give us a hundred different kinds of fragrances, vying with one another in delicacy and uniqueness. Only the *scent* of the noble rose is a phenomenon for me.

On the other hand, the new range of forms, and even more of the colors white and yellow, and of course red especially, is not insignificant. With this means one can create truly varied garden scenes. In groups on terraces, on walls, by the water, in large areas, and in open bands; for every purpose there is a variety that is particularly suitable. Thus, for planting roses, the often emphasized rule that varieties and colors should not be frivolously squandered, is particularly valid here. A gaudy rose bed is dreadful. The Läuger rose garden at the Mannheim Gardening Exhibition provided a beautiful example of everything that might be achieved with roses, while his counterpart showed a perfect contrasting example, of how the "royal rose" can have just as banal an effect as any other flower.[36] Although it is of course better to have fine roses together with other flowers, especially in the small garden, it is indeed an established fact that roses are best among their own kind. Only a few flowers, like lilies, gladiolas, clematis, reseda, and perhaps carnations, and for remontant roses perhaps autumn asters, will truly complement or improve the effect. Among this many-branched group of flowers, related characteristics are felt not only in the sections, species, and subspecies, but perhaps even more in the varieties, as a cultivated taste may easily discern. For example, the fiery red hybrid tea *"Gruß an Teplitz"* in respect to all its qualities goes extraordinarily well with the white remontant *Frau Karl Druschki*, but the droll little poliantha *Ännchen Müller* is very poor as a border to a bed of long-flowering tea roses.

Not least, the usual means of growing roses in our obligatory rose gardens has resulted in a generally dull expression. It is particularly the *high-stem form* of the rose, of one-and-a-half to two meters and more in height, that continues to thwart the potential expression of this garden form. High-stem roses today are literally a garden pest. If we could at once cast all of the "little high stems" [*"Hochstämmchen"*] out of our gardens, then we would be astonished at how many tolerable gardens Germany possesses. For the reasonable requirement of having roses closer to the eye, we have long had the so-called half-stem of one-half to one meter in height. With its help, we can set up gradual changes in height without spoiling the spatial effect demanded by the eye-level viewing requirement. One can confirm the generally drastic, demoralizing rhythmic effect of high-stem roses in our so-called public rosariums, so here I have included an illustration showing how these can be judiciously and well composed.

Those who are determined to "wander among the roses" have the truly lovely *climbing roses* for this purpose, which have not been given enough attention. We can train them in the form of weeping roses, rose standards, in garlands, on arches, or on trellises and on the pergola. In the latter case, whenever possible I would choose an open form, or certainly only a simple one, or a trellis frame open to the top, to allow for blooms on the interior as well, that is, toward the usual standpoint of the viewer.

The climbers that are most closely related to these and that partly overlap with them, are the increasingly popular *wild roses*. I know people who are by no means uncouth, who wholly buy into our entire business of "refinement," rather than the inexpressible noble freedom of a wild rosebush. But there is scarcely anything in the entire reign of flowers that can parallel the sheer, sensuous lushness and the gracious elegance of a *multiflora rosebush*. And our *cinnamon rose*, *cabbage rose*, the *sweetbriar rose*, the *Austrian copper rose*, and *rugosa rose*, the antique *centifolia*, the *villosa*, and the *moss rose*, and even the common *dog rose*: each species is a truly distinctive, royal beauty in itself!

What more is there yet to say about roses. Whether the modest luster of the simple ones, or the robust lasting quality of the double ones, or the buds here, or the full blooms there are to be preferred, and more besides. However, these pages are not for a professional handbook. Only one more thing is to be wished for with roses, apparently a paradox: *that there were fewer of them.* When today some element in the garden is to be floristically emphasized, then the odds are a thousand to one that the rose will be chosen. The whole carnival of varieties is the result of a generation of arbitrary amateurism and a fantastic collecting mania grown grotesque, which can no longer be enhanced to advantage. One will understand this when I cite the running entry number in the rose catalog lying before me, which doesn't even conclude with the figure of eight thousand. Eight thousand rose varieties!

Are there no other flowers? We should set up a commission, a proper German "comité," with the task of establishing a means of bringing an end to this calcified and pointless fashion absurdity within the kingdom of flowers. At the moment, we have far too many rose gardens—and at that, too few are actually beautiful. The one among us who frees us from our current mad chase after the noble rose, will in truth win back the rose for the garden.

It becomes clear of itself, that if we want to plant a number of these flower garden forms at the same time in one single garden, then they will require a definite relationship to one another. Yet our enjoyment of this garden creation can indeed be greatly increased, when from the beginning we consider their relative location, size, primary blooming period, and above all their different characters, in order to

Rose park, Berlin-Britz
Joseph Kumpan

TRANSLATOR'S NOTE
Josef Kumpan (1885–1960) worked with Migge ca. 1913, and later became a successful garden designer in Prague. See Gröning and Wolschke-Buhlmann, *Grüne Biographien*, 209.

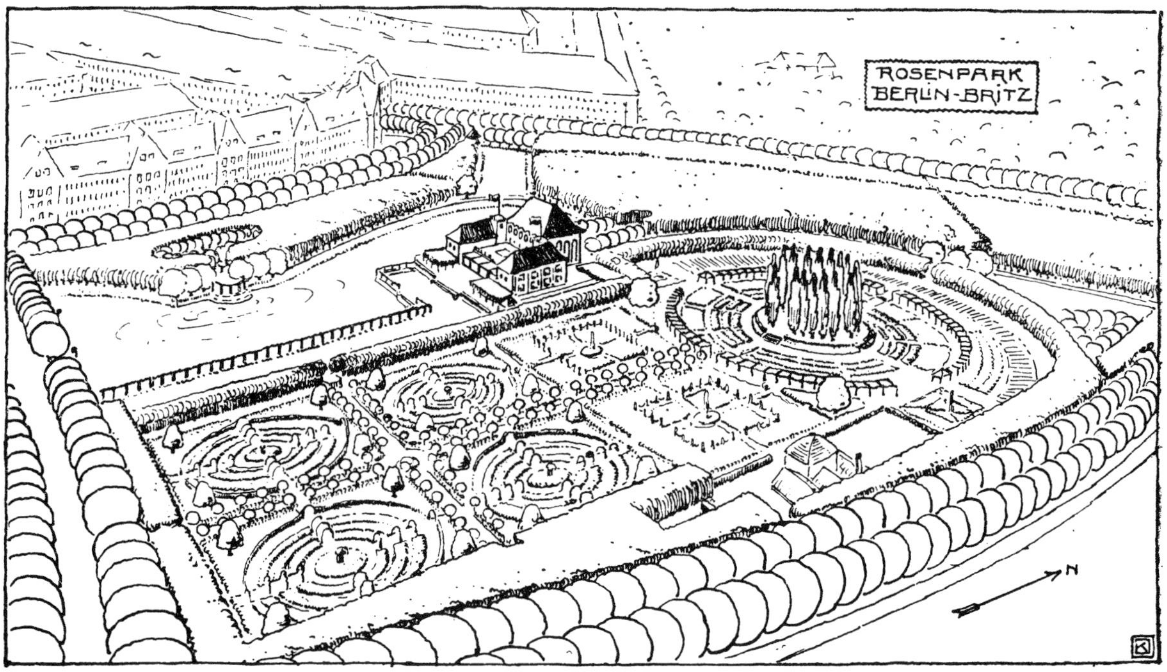

organize the different flower gardens into a *system*. In addition, the paths connecting them to one another and with the house can then be even more well laid out.

I know of a shady elevated promenade bordering a large lawn. At its base—the lawn to the left, colors to the right—lie all of the flower and rose gardens that we have touched upon here, and yet a few more, all in a well-ordered whole. Is that not a truly delightful avenue?

C. Nuggets from Practice[37]

A good garden will always make a relatively luxuriant impression. This sense of external perfection requires that the plant materials and the number selected not be scrimped. Most of our contemporary gardens suffer from too pedantically precalculated vegetation. The "honored clients" should finally show some appreciation for the superhuman [*übermenschlich*] origin of garden materials, and show some understanding of it. Plants are not building stones that can be numbered to the last one. As carefully as I tend to plan my projects in advance—something is always lacking, and I have always had the greatest doubts about gardens where the cost estimate left no room for adjustments, whether it was high or low. Just as all confident garden eras employed *all* of their means luxuriantly, meaning vegetatively, so should we. *We must above all strive to attain an appropriate garden expression as our main priority, and preferably realize our conceived garden image as closely as possible.*

Naturally, it is not the same when I have to create a garden in the region of the Mark, in the loamy soil of the marshes of lower Germany, or in the deepest humus of a riverside meadow.[38] Yet all the obstacles that the *soil* often presents to the garden are nearly always overcome through ever-improving technology, with vigor, and above all with patience. Then will "many weeds and seeds"[39] also sprout out of stones, which one thought impossible before. In the clay and moor-like western slopes of the Harz Mountains, for example, gardens thrive exceedingly well when the soil is thoroughly worked and loosened. I have even helped lay out gardens among the porphyritic stone of the Saal Valley. It is as though the vegetation there had an understanding of the hard nature of minerals. Not only may the vegetation in this craggy sterility appear as rich as that on the pleasant plains a few kilometers away, it is also more than satisfied here. Twenty to thirty centimeters of humus is sufficient to support the most lush green areas, and no one suspects that lindens and chestnuts as high as a house, stand on a layer of soil of no more than three-quarters of a meter above the local stone, and often less, and still thrive wonderfully. On further investigation, one may also notice that their roots have deeply penetrated the rocky debris, which was first loosened with a hoe and pick.[40] Here lies the secret of nature: many stones, as well as sands, are not completely lacking in nutrients for plants, which they know how to extract with the help of root hairs and their own internal bacteria. We must only prepare the way for them.

One speaks so often of *plant-hostile soils*. But sterile silt or clay, for example, can be supplemented with sand or dung containing straw, and thus be truly reborn, and in the case of the finest sand, humus, cow dung, or weathered clay also help. Damp "sour" soils should be drained and limed, and all of these in combination take deep freezing in winter well. In short, we have every kind of aid to realize our garden ideas, however much mother earth may rebel. The key to success in the garden is above all the correct and thorough preparation of its soil. The measures here are naturally varied, but even the finest flowing sand must be worked, that is, dug and plowed. All too often there is a lack of understanding and scrutiny here. Mechanical soil preparation, in connection with corresponding, i.e., with *different* kinds of appropriate animal and chemical fertilizer in respect to different plant needs, have the deepest influence on the lasting growth of the garden.

Almost as critical as the characteristics of soil types, are the varying *ground water conditions*. Here extremes on both ends are particularly disadvantageous. Most plants suffer from dryness when the groundwater lies too deep, and they sicken and weaken when their roots are in standing water. Such soils will be cold and unproductive. About five years ago, I laid out a garden on a boggy field, with the necessary fill material gained from ponds created at the same time.[41] In spite of extensive drainage and the application of every imaginable chemical and physical improvement to the sterile mud, as far as the vegetation, this didn't succeed in "getting it going" for a long time. Only now may one speak of the lush abundant growth and fully realized planting concepts, which guided me at the time. It was the then unchecked groundwater that hindered rapid aeration of the soil, and thus plant growth as well.

Another strong influence on the specific expression of the garden is the *climate*. The mild, damp sea climate of Hamburg or Bremen, the warm sunny cliffs of the Taunus, provide room for varied plant potentials that locations on the interior at the same latitude don't have. Even more decisive is the *altitude*. In this case, a few hundred meters more or less above sea level indeed play a role. In the brisk air of the Bavarian highlands, or even the wastelands along the Rhône, or the Eichsfeld, it is best to be modest when starting a garden. Nevertheless, based on my own experience I won't speak of insurmountable obstacles to the creation of lush and varied garden designs, when one may cleverly adapt to given conditions. However, I fundamentally don't recognize, for example, that a piedmont garden is a type based on local necessity. Of course that doesn't mean that I underestimate the potential unique beauty of the "heath garden" as described by Lichtwark. But just as with adaptation, stimulating *contrast* can also be emphasized, here with the heath garden motif. This will certainly occur when a garden provided with all the

plate 16

Phlox and saxifrage (beds), rugosas, roses, and Taxus pyramids have a robust bearing, providing pleasant vegetation.

PLATE 16

comforts expected by an urbanite shines forth like the sparkling light in a painting, as a fresh oasis within the austere monotony of the wide heath. And *this* pleasure of the owner partaking in such a garden is certainly not minimal, namely, to be reminded every day: of having *wrested something so unusual* from reluctant nature in this place. When a garden is established through clear distinction from its natural surroundings, we strive to evoke this through the selection of motifs from the beginning where possible, preferably through contrasts rather than unquestioning subordination. It is superfluous to introduce a simple garden pond in a lakeside landscape, and it is a denial of all deeper, rhythmic economy and of all understanding of psychological principles, to build a rock garden in the mountains. Pleasures, especially in the garden, are relative.

With all this I wanted to say only one thing: if we like, there exists no place in Europe nor in countries without great climatic fluctuations, that prevents the full application of modern horticultural methods in the spirit of these pages.

Regarding garden locations, we only have grounds for resignation with the urban garden in the strictest sense, and then only minimally. Here indeed, "only in limitation does the master distinguish himself."[42] Cities, particularly metropolises and industrial areas, fully exclude half of the entire range of material, when the term "garden" is defined by a certain degree of health for all of the individual plants contained within. If this situation were considered more carefully, then numerous garden complaints of which ninety-nine out of a hundred understandably originate in the cities, would no longer be made.

Thus, it is not so much landscape and climate that influence the appearance of a garden in the first instance, but rather the *labor of humans*. What is decisive in the end is: *accounting for the more restricted situation*, particularly the character of the terrain and the prevailing wind. Vigorous protection of the garden through walls or dense plantings at the exposed boundaries, together with intelligent exploitation of morning and afternoon sun, immediately strengthens vegetal potential, that is, the quality of the vegetation. I must point out that: technical control over gardening problems (on a rhythmic basis) is based on good solutions to the three primary questions: *soil*, *protection*, and *sun*.

Those are a few of the preconditions necessary for vegetation that determine the value of the garden. But there have always been people who simply can't wait. They want at least the structure of the garden to be ready immediately. Admittedly though: if in ancient times emperors and kings commanded extensive gardens to grow up in dry deserts, magically bringing forth a paradise from the ground using primitive means, then our own overly clever age should be able to do no less—when required. Getting the best of nature is a central aim in garden making, and perhaps the only real justification for the work of professional gardeners.

Our technology today has also improved for the garden, to the extent that when prominent effects with vegetation are immediately desired for structure and shade, *stronger trees* of up to three-quarters of a meter in cross section may be planted. The suitability of different species for this purpose varies greatly. Oaks and beeches are difficult to transplant when older, while birches and hawthorns are almost impossible. By contrast, locusts with their limited root systems do relatively well. It is even easier with *maples, ashes,* and *fruit trees*. Here *chestnuts, lindens,* and *elms* are also particularly suitable. *Conifers, rhododendron,* and all plants that hold root balls, as well as *poplars* and *willows,* have almost no limitations. However, two to three years must pass until these recover from the procedure, especially for older specimens, but once they begin to bud normally again, decades have indeed been gained. If our garden area is completely lacking in trees and the means permit, then we should always employ this aid. The advantages for the use of the garden, for plant growth within it, and for its general appearance are indeed extraordinary.

Our *shrub species* are generally more sensitive to transplanting when older. They each react very differently in any case. A *hornbeam* hedge is already very sensitive at five to six years, yet I was able to transplant a very old *laburnum* of about thirty centimeters in cross section, which today continues to blossom beautifully every year. But in the case of shrubbery, where the advantage is not so obvious, the demand for stronger plants is not so great.

By contrast, hardy *perennials* in the garden should be thoroughly dug up again, if we are not to wait for years for them to bloom. Here the species, along with soil and fertilizer, plays an important role. Above all, here one must certainly think of the necessity of good individual preparation of the subsoil. Even the strongest plants grow poorly if they are stuck in a soil that is sterile for them—and all untreated soils are sterile to a certain degree.

Occasionally, "good" or "bad" in terms of soil conditions is to be interpreted differently, even for the same plants. On damp ground, a *swamp cypress* grows naturally, that is, in a narrow pyramid of almost pedantically good proportions, while on relatively dry, water-bearing sand, it doesn't grow poorly, but rather picturesquely like a pine in old age. *Birches, spruces, oaks,* and other trees are even better known for such occasional extreme tendencies. When newly planted, these trees grow characteristically in all conditions, only showing adaptation to location in their *later* form. Plants are beings like people, which although of many variations and of chance birth, all owe their being to *one* mother, the earth.

The technical composition of plant materials in our gardens has received much too little attention. This escapes the judgment of laypeople from the beginning more than any other factor. In this respect particularly, good *root systems* of plants are to be sought, the result of *persistently good* treatment of individual specimens in the nursery. The future development of a young plant entirely depends upon a proper, i.e., characteristic root capacity, in undamaged form at the time of planting. For older specimens, the natural preservation of the crown plays a greater role.

However, what good is the most careful observation of all these requirements, if we err in one: namely, the *season*. A great deal depends upon which time of year we put the plant in the ground. In this regard, current superficial practices cause the greatest damage in the garden. An old debate follows, whether autumn or spring is best for planting. Autumn planting, particularly when early, according to my experience is definitely the most advantageous for woody species. A strong winter can cause damage, particularly when it follows a mild or damp autumn. An especially mild winter even more so. Another influence here is often an extreme between summer and winter. But whatever the cause, limited losses are always easily replaced, and when planting in autumn the great advantage remains so that in spring the plants are already "established" and thus in the first summer develop much better. Another consideration is that the necessary intensification of garden work in the spring results in a notorious workload for the gardener, certainly not conducive to the careful handling of individual plants, indeed often making it impossible, giving the advantage to *autumn as the primary planting time* without further question.

For some plant categories, experience has proved some exceptions. That the *arborvitae* and fir species are best transplanted in April/May is well known. I also propose that whenever possible, most short-shooted trees, such as *beeches, oaks, birches,* and others, as well as a few particular perennial species, should be planted in spring, especially when the soil is heavy and damp and the site is rough. However, in lighter soils, these are best planted in autumn.

Above all, the danger of *late planting* is greatest in spring. Not only because the changeable weather during this season tends to thwart all planning, but also because of the all too common occurrence of not having enough remaining time to finish supporting structures, which were started too late to allow for the normal realization of the surrounding garden area. This results in all kinds of stopgap planting measures, the reason for these being unrecognized or in any case not admitted, much to the annoyance of all of those involved for years. For example, I believe that the awkwardness shown in the greening of house walls in Germany, in contrast to other countries, not least is generally due to the utter impossibility of the building construction worker appropriately handling the vulnerable vines at the very last. Ensuring a timely planting must therefore be an important task for the owner, when the building contractor fails this living material in the garden out of lack of interest. In order to avoid this difficulty altogether, I recommend that for larger projects and where circumstances permit, the majority of the garden should be realized *before* house construction. *Plan in the summer, plant in the autumn, and refine in the spring*: that is the good standard for good garden making.

plate 17

A Scottish sweetbriar rose grows over a seating area; in the middle, a vase on a stone plinth.

PLATE 17

And one more thing: even when all of these problems can be dealt with through good preparation, the final result can still be jeopardized when the later *maintenance* of the garden is neglected in whole or in part. And here it is indeed insufficient to observe only technical quality, practical knowledge, and the thoroughness of the maintenance, for the overall design must not be overlooked: the spirit of the garden creator must also watch over its further development. And this takes time. The widely held opinion that the newly laid out garden is "done" just as soon as it is handed over for use, cannot be contradicted strongly enough. Thus, it follows that, *for a newly laid out garden to be properly brought to its completed state, the advice of the one who conceived and realized it is indispensable in regards to its maintenance, in whatever form and over a given period of time.*

Unfortunately out of generally petty, as well as short-sighted and therefore in the end uneconomic calculations, many truly well-planned gardens are often spoiled through the means of their further maintenance.

VI

GARDEN STRUCTURES

As a cultural object our garden wants to be enjoyed. Yet only through opportunities for casual lingering and gazing can it be appreciated in its full worth. A garden without seats and resting places scarcely deserves its name, and uncomfortable, rough, or artificially imposed furnishings are its ruin in the end. Built structures in the garden must also grow with and out of it. The garden and its fittings must be *one* organism.

A. Garden Furniture

The simple garden bench, a seat, and a table have always been the most important things for the more intimate use of the garden. It is a fact that these first appeared when the gardens of a people emerged from an embryonic state. And it is just as interesting to observe clearly that on occasion gardens in a given period were still of minor beauty and of no exemplary value for others, as long as the equipment they contained was technically poor, unrefined, and in the strictest sense, unusable. There have never been good gardens without good furniture.

Accordingly, when we rely upon garden furniture to support garden living today it must be problem-free in all respects.

There is much to consider even in its manufacture: garden furniture stands out in the open in all weather. Therefore, good water runoff and general durability will be considered in terms of the slope of flat surfaces and the general avoidance of large wooden elements. For the same reasons, a garden table should not be glued or veneered, like a piece of interior furniture, but instead nailed, screwed, and pegged. The wood must be able to "move." This is a special problem in the manufacture of garden furnishings. The cabinetmaker is too exacting for this purpose, and the rough carpenter and even the finish carpenter are generally too crude. Garden living everywhere has its own requirements and these require special treatment here as well.

The construction of garden furniture depends upon a certain degree of *portability,* as required. Its garden table, and in particular benches and chairs must be able to be moved at will, today here, tomorrow there, when one wants to fully experience the charms of the garden. Average garden furniture thus must be relatively light, but nevertheless stable.

Furniture planned for a particular location can be handled somewhat more freely, with a three or more cornered, semicircular, or circular bench, all conceived as seating for more than three people. Placed in niches within hedges, on terraces, and under individual trees, as "built-in garden furniture" these can be much heavier, with more freely designed forms.

The *material* is also not unimportant. We give preference today to tradition-honored wood, as opposed to solid, but cold and bulky iron. Combinations of wood and iron, as suggested by the properties of both materials, are still uncommon. I recommend lighter softwood for general purposes, but heavy oak, birch, or pitch pine for load-bearing elements, for round forms, armrests, frames, latticework, and heavy pegging. All wood to be utilized should be dry and well chosen and protected.

The desired solidity of construction and material also correctly influences the *form* of furniture. First and foremost, this should take functionality into consideration. Today's garden furnishings are generally not designed to fit the people they serve, but rather for pleasant decorative effect. As much as the emphasis on the decorative may be necessary for certain requirements, it is nevertheless impractical to wholly or partly ignore practical intentions for our garden furniture in our contemporary functional green spaces. Just as with the customary development of useful and really good household equipment, such furniture cannot simply be designed or "drawn up," rather it must be progressively "developed" starting with the original sketch, followed by multiyear testing periods in the garden and one design drawing after another. The examples illustrated along with many others were produced in this way.

Such a process may at first lead to a certain austerity. But there is a means of compensating for the cold appearance of our garden furnishings during this necessary development period. This is *color.* The obligatory garden bench is indeed

white. "Painted white" has become the fashion of our day. This isn't really of our era, rather we have borrowed it and its decorative effect from previous periods. But over time, a more refined reaction has arisen in opposition to uniform, dead pure white in the garden. Our newer, intimately designed house garden often has, more than we are aware, situations with a pronounced domestic character, which also demand modest integration of the constructed elements of the garden as well. Just as in our efforts with the interior, here it becomes apparent that there is also a strong need on the exterior for mediating, harmoniously contrasting colors, much more appropriate for this application than white. One may imagine how well a gray-colored bench with ocher yellow or red armrests complements the dark green of a Taxus hedge, or how a deep blue seat accented with bright white or red is effectively set off by a gravel ground treatment. And even if we are to use white paint on our garden furniture in various situations to advantage, then relief through gray or ivory colors is of value for the overall impression and particularly for practical use. Garden furniture like other garden architecture, often takes on an unexpected liveliness through color, and the ornamental effect is unmistakably improved by a simple stencil here and there. In general, I am against the muting of pure color tones in the garden, as is common in living rooms. If I use colors here, then I do so with conviction.

B. Arbors, Pavilions, and Pergolas

These are some of the most common names for a variety of protected resting places or walkways, of wood, stone, or iron, in combination with vegetation. Although sitting areas enclosed by hedges (beeches, lindens, Taxus, arborvitae, and many other short-shooted shrubs) are easy to make, this usually takes too long; green walls take time to reach completion. Everyone wants to be able to use their garden as soon as possible. Thus, people turn to aids: one erects a light-sheltering latticework and covers it with vines, or builds a more or less permanent arbor.

It has always been so, as long as there have been gardens. Only the forms of these garden expressions have regularly changed. Continuing through the so-called Biedermeier era,[43] "the" garden pavilion was a little house with a permanent

roof, with a door and windows, erected in the corner or in the middle of the garden. It had to be *enclosed* because it offered protection from quickly changing weather, and guaranteed a retreat when the garden lay away from the house, which was often the case. Its form originated with the reasonable requirement of making the enjoyment of the garden possible under any circumstance. By the standards of the time it was a *given*.

However, today we have a much more developed domestic culture. The verandas, balconies, loggias, bay windows, and terraces of our buildings assume many of the functions once fulfilled by enclosed pavilions. We are also generally people who are more fond of sun and fresh air, and have thus become more "hygienic." We love sitting outside among the green and flowers, more than being in enclosed rooms. The pavilion is thus more inherently suited to larger park areas. I only use the garden house for gardens in the true sense, if it all, in a fully or partly *half-open* form.

Half open! This is, I believe, a principle that we should follow when erecting small structures in our garden, if we don't want to set up mere showy displays. This principle emerged of itself—from use—and it is applicable for all sizes and programs.

plate 18

Two practical as well as original garden benches, designed by my former coworker G. Ammann, executed by Froebel's Erben, Zürich. Here as well, a sparing use of colored paint.

TRANSLATOR'S NOTE

Gustav Ammann (1885–1955) worked with Migge at the Ochs firm early in his career. He returned to Switzerland in 1911, and worked for the renowned Zurich garden design firm of Froebel's Erben. Ammann later became an important independent landscape architect in Switzerland. See Johannes Stoffler, *Gustav Ammann: Landschaften der Moderne in der Schweiz* (Zurich: gta-Verlag, 2008).

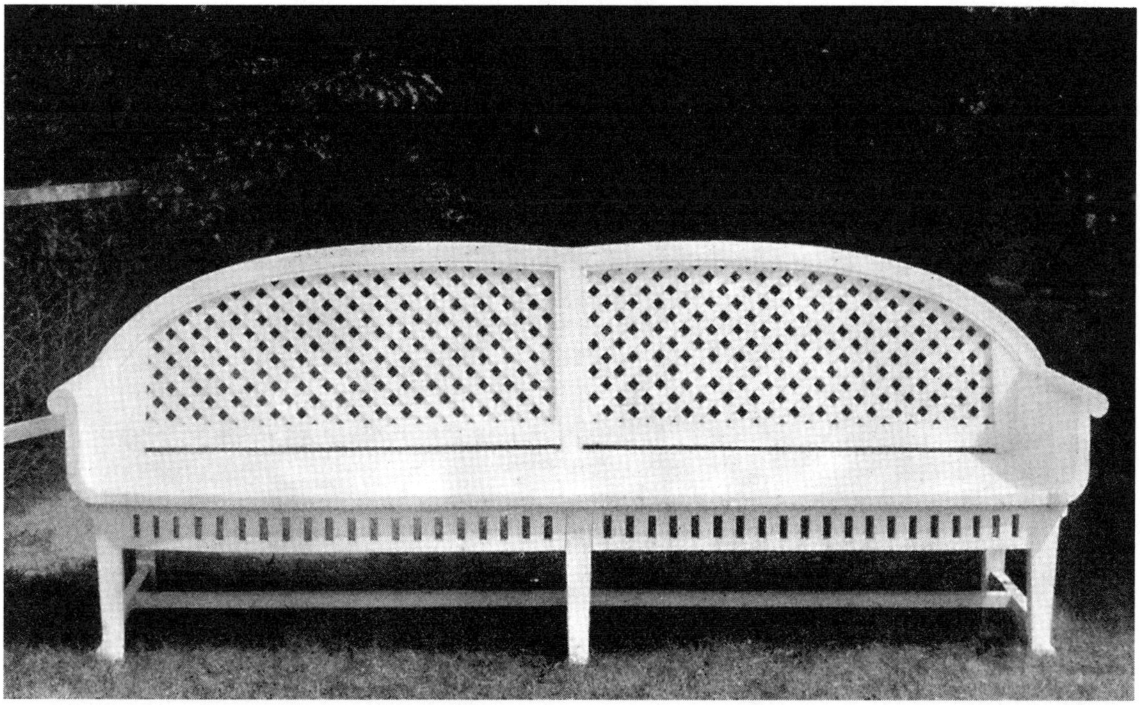

Thus, if a client approaches me in a particular way: "I would like to have a seat built in my garden: with protection from wind and direct sun, but still airy, with space for six to eight people, where it is possible to place a table and chairs, for a price of so-and-so many Marks."—thus follows almost automatically a type that is similar to those shown in the illustrations here, depending on whether a permanent roof or only latticework is desired.

Latticework above one's head is generally an uncomfortable thing. For a pergola, through which one passes, it's fine. But sitting under the thick growth of vines for any length of time is not for everyone. In spite of this, there are those by nature (the author included), who for the beautiful experience of sitting under a fragrant rose, are also heroically prepared for the possibility of being beneath the productions of the spiders that sometimes crawl down your neck.

Just as with the roof, the *floor* must also be considered. In a garden pavilion, this must not be "cold to the feet." In cold areas or under permanent shade, a raised floor should thus be planned for. A layer of clinkers on a simple sand bed, edged with brick, is suited to this purpose and beautiful in effect. Such a "screed" is cheap, lasting, easy to keep clean, and when a pattern of stones is laid out within it can have a very ornamental effect. Naturally any kind of paving stone (cement, sandstone, ceramic) also fulfills this purpose.

But in the end even the most charming pavilion can have an embarrassing effect—when it falls apart. It must be sensibly and appropriately *constructed,* meaning, for its size, purpose, and material. What has been realized in this respect within so-called garden architecture today is not to be believed. The most elementary principles of statics that one should be able to master by feeling, are disregarded. As ever, lack of expertise is vindicated through the most audacious measures possible. Has one already noticed this? The most insignificant garden artists regularly build the most uncouth pleasure pavilions.

The construction is dependent upon the material that we choose. Wood is again the best for all of the smaller solutions. For the simplest needs, for small pergolas, arbors with open roofs, etc., one may revert to the use of log timber (without the bark!) that can be used in right-angled (not naturalistic!) compositions to realize appropriate garden structures that are immediately favored by many vines. But

in general planed and painted woodwork is the best for our more refined domestic sensibilities. It is also scarcely more expensive. Studs and laths are now industrially produced in great quantities, and if nailed together then laypeople can often easily assemble garden architecture with them. The workmen's and allotment gardens of our big cities offer abundant examples for study in this regard.

For larger pavilions and garden houses, *masonry* is the correct material because it is stronger. It can be of exposed brick or plastered—when it doesn't act as the foundation—built as half-height walls, or in the form of columns carrying the roof, or as more or less open walls. Three such walls in most cases are sufficient, with the fourth side open to the sun.

In the garden, *concrete* and *iron* are less tried materials. These are still foreign to our eye in the garden. Some people are wholly unconvinced. But following my own experience, these materials stand well among green and flowers when properly formed; indeed, even harsh concrete can be so fancifully formed that it appears to perfectly complement the garden.

The *roof* still remains. In addition to reed, straw, and wood, we also have more permanent roofing materials, such as pan tiles, fish-scale tiles, and slates. Ruberoid,[44] and when necessary roofing paper, may also be used. Anything may be used that is technically appropriate, provided that it can be paid for. But "it" is often not paid for. Even garden seats of all kinds must be carefully calculated. We must never forget: *today we create more than a hundred times more gardens than before!* Most of these are not very big nor burdened with extravagant debt, and even for the smallest ones there is a serious lack of applicable, here meaning affordable, models for arbors and covered seats.

But this economizing, as imposed upon us to a certain degree, has still had all kinds of good results. It leads to simplicity, and restrains inappropriate desires for ornament. It's the aestheticizing arty trends that have suppressed so many good forms within our daily environment, which otherwise would have appeared. Useful things of an elementary nature, such as garden pavilions, are indeed beautiful when of good quality.

The same is true for the frameworks that we build of wood or iron as support and background for our various climbing vines. An old garden structure that was

often used in oriental gardens especially, can practically serve the purpose of providing partial shade for promenades. A *pergola* is airier and lighter than the often dank and dark arbor walk. But because the necessary framework of strong material always dominates the surrounding area to a certain degree, a pergola should never be arbitrarily erected in the garden as still often happens today. This construction should not be thought of as a thing in itself, instead it needs to have an organic connection with the house, or act as a significant structural focal point for the site. Pergolas best fulfill their function when they provide a direct connection between house and pavilion, or to another important seating area within the garden. When beautifully constructed, covered over with a woven texture of plants and flowers, and given a ground cover of colored gravel, then it is truly a fairer, greener corridor.

C. Small-Scale Architecture in the Garden

This is a field of work in green that is as varied as it has been unappreciated up to now. It begins with the beanpoles or iron stakes that we use as supports or frames for high beech hedges and linden arbors, and finally ends with the flight of steps leading from the terrace of the mansion to the garden. The stave that supports the rose belongs here, along with the arch that spans the path, the street fence, as well as starling houses or the hedges that enclose the chicken pen, the flowerpot, or the fountain that delights our eye with the essence and beauty of water in so many ways.

The paving of paths, the construction and material of walls that support soil and plants—these and much more will be considered here. All of these are building aids for the garden. But in their forms and their technical execution—have they been taken seriously somewhere? By no means, as scarcely as their existence is more generally and properly understood at all.

In earlier garden periods, this was not the case. Then one knew exactly how much the rapid, full completion of the garden design depended upon the useful structure of a pergola, and one considered well-proportioned, beautiful, comfortable stairs and steps to be absolutely indispensable to the unhindered enjoyment of the garden. Nothing is perhaps more characteristic of the wild or better, primitive,

state of our garden today than the usual *garden steps*. Often patched together from the most illogical materials such as log timber or rubble, it is particularly the height of the steps and their random scale that every human body type and all experience prove to be ridiculous. The landscape garden period of the nineteenth century utterly ruined these, so that even today otherwise capable architects are unable to free themselves from the usual neck-breaking external house stairs. In my gardens, I only use steps that exceed fifteen centimeters in height when required by necessity, but in a level garden I often go down to a nine or ten centimeters rise. It has been a long time since one spoke of a culture of stairs in itself, such as the finest feelings and most monumental dispositions were once capable of, but it is certain that in the future we cannot realize any worthy garden design on uneven sites without well-proportioned stair layouts that are good in every respect.

These observations may be applied at will to almost all small-scale garden architecture. Here much has been neglected. Thus, for all the varied garden structures of wood, iron, and stone, every commonly accepted rule of good construction and of so-called material appropriateness is here more important than ever. It is especially important to ensure that these are subordinated to the prevailing material of the garden, namely the plants: garden architecture, particularly the many incidental small structures, should be *integrated* and *subdued*, not, as still often happens today, overwhelm and suffocate the overall garden image, particularly when it's young and underdeveloped. Thus, a beech arbor, which of necessity could grow on its own, must not be designed with massive beams painted with the brightest possible white, when in fact iron straps or bars with wire, or a light wooden latticework would be perfectly sufficient. And in the same way, it is a not unimportant misunderstanding of yet a small functional detail, when the supports for future rose columns, a sweet pea pyramid, or of a lattice partition in a flower garden, are made of cyclopean proportions requiring years of expense for the garden design. It is a truly silly practice, when today in our gardens rose stems are typically planted in herds on thick, crude, loud, white square sticks, advertising modest blooms like small crowns, with kettledrums and trumpets. Within the tactful balancing of external elements and plants, many finer harmonies lie hidden.

In the end, when other means are lacking much can be softened and brought into unison through color. A white climbing rose that simply can't be fully appreciated on a white trellis, stands out in all its beauty on a blue-gray or brown background. It would be too much to introduce and discuss all of the specifics of this area in detail. Here, in the most flexible area of garden techniques new suggestions are now being constantly made and new materials introduced, which is absolutely indispensable for the design of small-scale architecture of a predominately ornamental character (fountains, urns, etc.). This belongs to the field of applied arts.

D. Garden Sculpture

The field of applied arts is not far removed from the old accustomed garden sculpture, as the only one among the so-called fine arts that can seriously contribute to garden design. And this would certainly include to advantage that which has grown out of the ornamental independently of architecture, that is, three-dimensional or figural sculpture.

What does it mean then to use good sculpture in our gardens?

In former times, an ornamental object like the garden itself represented great people and their refined sense of beauty; it was the concrete expression of the fair

plate 19

A garden house on a wall made from rough-hewn stone, with amusing decorative painting. Designed by G. Ammann. Executed by Froebel's Erben, Zürich.

PLATE 19

social life of a closed circle. The intellectual tendencies of the past favored models from classical mythology, harmless personifications and similar things. Our social life is different. The spirit of a person who feels the age of invention and natural science in every fiber of his being, and who also has a different psychic nature and life rhythm, requires a particular artistic expression of their desires. This new individual absolutely cannot be served with discreet allegories, harmless nymphs, centaurs, and philosophical idylls. Our artists generally don't consider this sufficiently when they complain of lack of interest and investment.

The special quality of this particular world of ideas is supported by a further dependent factor: the *material*. Not all materials are of use in the garden. For this kind of location and owner, marble is disconcertingly pretentious; it is also often excluded because of climate. Sandstone fits well within the garden organism, and limestone and shell limestone are even better. Yet all of the materials that facilitate reproduction are given too little attention in respect to contemporary garden ornament. In this case, I am thinking less of bronze because of the limited possibilities of exhibiting it in the garden (it needs a bright background), rather much more of weather-resistant majolica as well as terracotta, and—of the many varieties of artificial stone. In addition to their low cost, these materials have the advantage of accepting artistic color treatments. In this respect, artificial stone in particular has a certain future. Yet our individualistic artists continue to resist this; but if they would only seriously occupy themselves with the special worth of such materials, then appropriate, mature forms would appear on their own.

At the same time, I submit that I don't hope for anything less from a new garden culture than an improvement of sculpture itself—although one can scarcely speak of a flowering today. In my opinion, contemporary sculptors on average lack the capability and conviction for the necessity: to see and create sculpture for its location, above all for a *space*. In recent times many, or indeed most of our sculptural achievements have suffered enormously from the indifferent manner in which they were exhibited. And, I believe, the mission is clear: in the open air, green and flowers—sculptural works should be created for people who once again find pleasure there (because there is no real art without leisure, and today's exhausted working

person prefers to relax in a beautiful garden), awakening instincts and thoughts in artists, which so far have only been initial concepts recognizable among a few. The causative effects of the simplification and increasing abstraction of sculptural art demand full expression, indeed, complete freedom from formal developments of a resistive or irrelevant nature. And directly for the sake of these characteristic, stabilizing qualities, our bustling era must set its mind on a renewed flowering of plastic art as a high goal. And I believe that in comparison to architecture, which sculpture naturally serves as handmaiden, here the garden itself can make even more of a contribution. Under its shadows, within its clear air and natural freedom, we may intelligently encourage a new era in sculpture to spring up.

The expert is all too aware that we are yet far from this goal, and that the foundation here is still quite insecure: nothing in my gardens has caused me more worry, work, and effort than the sculptures that simply appear to be there naturally, along with their hard masters and capricious creators.

From these discussions, we learn that although relatively ignored up to now, garden structures encompass a field of work that can span from the simple flower stake all the way to garden sculpture. (See Appendix 9.) This also has great spiritual significance. The simplest technical creation, perhaps a machine-made mass product of pronounced typical form, pushes the personal and unique creations of the artist aside. An image which can no better illuminate the inscrutability of life in the garden.

WAYS TOWARD NEW GARDEN ART

VII

A. Ach, We Have No Art!

One may take this as seriously as one likes: I have a certain prejudice against "artistic" activity within our contemporary society. In any case, in respect to the decisive influence of art in our society; here it *cannot* really *authentically* produce. Artists today appear to me, with few exceptions, to be romantic and isolated from the world, not only because they haven't yet found an inner connection to the events of our day as is virtually undeniable but also because they have scarcely shown a universal *will* toward this end. What our artists today generally strive for could have already been achieved by a Caspar David Friedrich or Philipp Otto Runge[45] in the middle of the nineteenth century. The ultimate fulfillment of the quests of a generation of artists depends not least on studying and shaping materials from within the sphere of activity of their own time. That is, to perform a task with the very *selection of materials* which others before them, however significant they may have been, simply could not have done. Does this in fact occur? In fact, very rarely does contemporary artistic practice as a whole base its materials in other than a purely superficial manner on the capitalistic-social or technical-scientific foundations of the nineteenth century; the cosmopolitan approaches of the twentieth century along with its moral intensifications and psychological refinements have scarcely found significant interpreters. The opening up of the universe, discoveries and inventions, the literal eruption of the technical era, our mass wars, the hope or the curse of capitalism among us, the new religious offshoots above all—are these insignificant things? Are these not the actual events that must incite our artistic wills to action? *Art*, i.e., our own and authentic art *is created* either *in connection with the ethical experiences* of our own time—or not at all. That which we try to describe as "high art" today is not the expression of a defined worldview, or certainly not of *our* worldview. In such a groundless state, an art can by no means become the aesthetic and moral leader of a nation. Over time this situation naturally leads to counter-reactions as well. That we so often hear seemingly justifiable complaints about the lack of the understanding of art and the need for it,

PLATE 20

is not least the result of the fact that still today we have not seriously tried to translate our actual lives [*Leben*] into artistic *experience* [Er *leben*].[46] Our art survives on its relationship to surrogates. One may say that a Johannes V. Jensen, Verhaeren, Dehmel,[47] Walt Whitman, and Maeterlinck, a Hodler, Meunier, a Strauß, Peter Behrens, and others, perhaps owe their true popularity and individual significance largely to the uniqueness of their creations, which followed the tendency indicated above—not simply because they were great individuals of importance to world history. For us, the question itself is the problem—*what* is art, and *who* is the artist.

The intolerability of this situation is indirectly revealed through the typical uncritical use of the term "art" today to describe each and every thing. If the noble purpose of art, its period of both highest achievement and freedom, and its strongest expressions, are only called forth through tentative understanding and agreement, then in brief: *to arouse enthusiasm* it is degraded everywhere. Our quaint folk art, our art society mania, museum art, the dozens of art-"isms" in the actual "business," our thoroughly organized art market, and above all the whole fuss about art in our daily lives—are we able to even slightly connect these with the concept of divine revelation, as did great men who spoke about and experienced this over the ages? I believe not.

Thus, contemporary art lacks authentic delights for its beholders. It leaves us cold. Only occasionally does it touch us in the artistic sense, but on the whole it doesn't really move us inwardly. A great pessimism grips all those honest and capable enough of serious self-criticism in opposition to the art industry that surrounds us. And we ask ourselves, if the old, well-worn paths don't lead to the goals, will yet other ways emerge for our race toward higher levels of life, *toward our own true art*. The people and the art of the current age are foreign to each other. Our art is dead or has not yet been born. Ach, we have no art!—

This brief declaration is based on the sum of impressions from my young life, and should be taken in that spirit. In any case, I naturally feel that I have neither the professional background nor the disposition to reveal this fine interweave in human life; the *generally uninspiring position of art* today is all that I wanted to briefly point out. And from the outset, I do this with the aid of this formidable universal human problem, to fend against all of the self-glorifying and special claims to art that our new field of gardening prefers to real work.

B. Art and Nature in the Garden

The translation of an idea into being, from credible concepts to tangible material, is a fundamentally different process in applied horticulture than in the otherwise

plate 20

Covered seat, half-open. Protected by old trees, and yet still with some sunshine.

closely related areas of architecture and interior design. Above all, the *possibilities* are more limited and less predictable. Plants as a material are too mutable and too unstable. They are *living*, and all living things elude human awareness and will. That which you imagine, dear enthusiast, your garden image, doesn't allow itself to be captured and fixed. It opens up and then shrivels in your hands, it suddenly changes colors and its entire dress. The pessimist in this case rightly asks: When is my garden done, when is the artwork, artist, finally complete? Is it in spring or in autumn, or even in snow and hoary frost, is it after five, ten, or twenty years, or when after generations, the trees murmur of eternal time? Ach, the garden is *never* finished! And thus there has never been a "garden artist" who was able to realize his ideas on the ground without distractions or impediments. Storm and lightning, drought and torrential rain, and above all time, only too often act as silent, uninvited coworkers. Most deserving of the name garden artist would be the one who understood how to most artistically handle such limitations. Thus what is important here, is how every garden image in every period has had in its most successful, most convincing elements *something unverifiable, unobjective, something illusory* in itself, so that one may often be able to say of a garden that it is just this special character that makes it so captivating. The quickest sketch of the painter or sculptor ultimately can be something rhythmic, an artistic act—but the garden flows, it constantly renews itself, it is beautiful and magnificent, perhaps as a divine creation, but it can nevermore be an artwork of human measure.

Let us not be too ambitious. So-called garden art is nothing more than a somewhat moody but nevertheless natural sister of "building and interior design," or better, of cultivated construction. I thus count it among the *applied arts*. As such, it shares the inner callings of all applied arts: being alternatively dependent upon purpose, situation, or material. Only by starting from this position, as "applied garden art," as contemporary horticulture, or however one defines the term, is it possible to facilitate the development of our future garden rhythms with a view to success. Here is the lever where human energies alone may be consciously exerted in order to renew the delicate concepts interwoven within the historical term "garden art." *Just as we have reached the same results from economical and ethical grounds, we must also acknowledge here that it is necessary to begin with the functions of the garden, in order to—perhaps—be able to end with its rhythm.*

Thus, all of the pointed struggles and demands for nature in the garden are negated from the beginning, meaning those that garden aesthetes, fantasists, and outsiders have always focused on. Up until today, this has merely succeeded in clouding heads and minimizing results.

The introduction of nature in the garden is rendered superfluous under the guise of *scientific plant biology*, that is, the study of life communities of plants and of the relationship of the locale to the economy [*Haushalt*] of plants. This system, namely, of trying to solve rhythmic questions through speculative means, suffocates itself. Thus, every kind of question eludes the direct influence of technique or science as a result of their nature: organizing reason and scientific understanding as the elementary principles of art are irreconcilable contradictions here. In a similar way, botany in itself is about as secondary for rhythmic garden design as the knowledge of concrete production is for the attractiveness of a house made from this material.

Just as unfounded is the concept—a translation and expansion of the previous—of seeking representations of *wholly unconstrained nature*. Simply because people need it (nature), as the child does the mother, so to speak. It should develop in the garden "as though on its own." But here also questions forcibly raise themselves: *When* should such nature be in the garden, and *how*? And above all, why then in the *garden*, when it does so much better outside of it, where it is furthermore accessible to all? To this and similar objections, a clear answer is still due.

As I have already discussed in the beginning, the desire for nature in the garden, allegedly in order to create a purely *autonomous artwork*, is superfluous in the end—because not really possible and also not necessary. *Impossible*, because of the unavoidable, illusion-destroying factors within the complex machinery that we call daily life, which experience has indeed shown to hinder the final potential artwork, and *unnecessary* here, because even though the garden is "only" applied [art] this still allows it to develop the blessings bestowed upon it in full measure. Then why all the effort! In any case, without anything further it is possible to incorporate purely natural plant groupings within the architectonic garden framework. In this respect, I can imagine relatively "naturalistic beds" of any size, similar to the way that I have tried to fulfill the desires of other devotees, such as dendrologists and botanists, in their own gardens.

For all purposes, the correct, meaning the sensuously and vigorously conceived architectonic garden principle, with its plant harmonies, its contrasts between uncontrolled and formed plant growth, and its constructed elements, evokes the greatest variety of charms imaginable here. All that remains, which to me is the finest and most valuable, may be quietly sought outside in free nature. But this measure of direct *contact with nature* that individuals justifiably want and must have, is also fully guaranteed within the architectonic garden. Thus, it also grows from its most tender, touching youth to the lush, full maturity of its years. Within, it also buds and blooms and holds the sweet joys of spring, just as it does the flaming colors of autumn. Only much more intensively, I think, with a much stronger impression on people as realized through a contented awareness of their own creations and inventions.

No, we contemporary people want nothing of garden naturalism. From whichever side we try to approach this problem, it evaporates in the face of worldly enterprises and the new field of perception within which we operate. There regularly occurs—a characteristic sign of our current garden rhetoric—a *circulus vitiosus* that moves from nature to science, from reason to art, and back again to nature in a never-ending cycle.

More than anything else, it is the *multitude of viewpoints* that still hinders an objective judgment of garden intentions. The modern person believes in everything or in nothing. To be truly human, i.e., to be culturally inclined, he would have to learn again to believe in *one thing*. What I believe in respect to the garden will be further established as follows.

C. Garden Types

In former times, there was a relatively limited number of people on the little speck of earth that is now called Germany. One could justifiably expect that when they endeavored to raise the cultural level of their lives, each individual would produce something of value. Every outstanding act was the accomplishment of a small circle.

However, now a *multitude* of people live within the *same* area, yet the needs of the individual have not really decreased. Matter, including spiritual matter, always remains essentially the same according to an old saying. Thus, this multitude of people in Germany, this mass, is hardly ethically obligated to produce more ideas and forms than those few. Certainly not in terms of numbers. Above all: they will be wholly *unable* to do this. The result of this is indeed logical, that is, now one and the same entity will always have to be equally valid for the many. *Such a thing, that is conceived for the many simultaneously, we call a type*. We mass people need types.

In terms of the garden, this is actually nothing new. Garden history shows us evidence that such types existed almost everywhere, at least in a somewhat limited and figurative sense. Spiritual and material influences stood in equal proportion in their production.

If one will, the utilitarian garden can be categorized as the first typical garden. All the garden art of earlier periods indirectly arose from this experience. Thus, the later ornamental gardens are nothing other than the spiritual and material elevation of the vegetable gardens, herb gardens, and orchards that preceded them. That there were definite forms of cloister, school, and botanical gardens has already been mentioned previously. But there were also *garden elements* that incorporated so much experience and appear to have found so much favor, that these were included as permanent features in the gardens of almost all peoples and times. Already in purely functional garden designs there are protected pavilions and promenades, with an external appearance that attempted to conform to the means and the social and moral sense of its builder, and even to the universal,

plate 21

A pavilion also need not always be enclosed. This airy pavilion stands above rose terraces, lakes, and green meadows.

PLATE 21

typical *majority*. In the Middle Ages, the so-called flower meadow, a lawn scattered with flowers with a fountain in the middle and rings of trees around it, was a common, consistently recurring garden element. I am also thinking of the medieval "orchard" [*Baumgarten*], that we again find as a shade-giving, fruit-bearing, or spiritually dedicated grove, among the Egyptians, Chinese, Persians, Greeks, Moors, Dutch, and others. I think further of the "parterre," the "orangery," and the "labyrinth" that come from French gardens. I think of arbor walkways and allées, of terrace forms as the ancient means of accommodating changes in level, of fountains, and of cascades as rhythmic representations of water, and much more.

And thus what matters here is that although gardens and garden elements appear to us now as having been born of one individual and seemingly spontaneously created, in fact, they represent the results of the garden desires of a *multitude* over a longer or shorter span of time. It is unreasonable and unhelpful for us to hope that the garden processes of our time may be of a different nature. On the contrary, if it is to be any generation, then it will be *ours* that must proceed from the generally valid unit to a well-proportioned whole. As thinking and self-confident people, we are certainly capable of attempting to condense general intellectual precipitates, as well as to formulate, propagate, and refine practical elementary principles.

When we review the field of our current garden living in this sense, we discover various things that refer to the development of typical garden forms. Not only have we embraced the arbors, allées, and groves, the pavilions and pergolas of earlier garden periods, and adapted them to our needs, in addition many new forms are also appearing. The *garden terrace* has already become a given element in our new *Landhaus* design. A *communal lawn* in front of the house, large enough to enable one to comfortably walk about and play leisurely games with the family, is becoming accepted. The *flower border* is returning to almost every garden as an ornamental element or as the source of cut flowers for the home. In new garden designs, we often see a separate little children's garden, connected with a play area for the little ones. Everywhere there are *play areas,* such as play lawns, running lanes, or gymnastic, tennis, and football areas, which have given some gardens today their special look, while in this era of hygienic self-discipline, *sunbathing* and *swimming* areas do even more. We note the psychological effects of the scientific and trade policies of our time in the *rock* and *water gardens* erected by the wealthy. All of the planting methods that I listed in the section on *vegetation* become new, typical individual garden forms through their contemporary treatment, and what I discussed earlier in the chapter on *Garden Structures* is also wholly related to the typical. One may say that everything designed for universal validity is good and useful, while on the other hand, nothing has the power to become typical if unhealthy at the core.

The repetition expressed by these garden elements today is by no means merely limited to the form of this or that unit. Thus, all those forms of *individual gardens,* which I introduced as the gardens of the metropolis, are nothing other than types, which came about under varied, yet unavoidably difficult conditions. However, the open gardens of the *Landhaus* beyond, or the urban park, are also subject to this principle. In the end, frequent use leads from free-form elements to defined combinations. A convention is established here in terms of the connection of garden elements to one another, their orientation to the sun, and above all their relationship to the house. An arbor walkway always forms a good functional boundary to the property, or between the utilitarian and ornamental gardens, while one orients a rose garden toward the morning sun, and as is practical the south wall of our house better supports ornamental vines than the north or west sides. The children's play area should be fully visible from the room where the housewife is, such as the kitchen; following an older established custom, flowers and evergreens should be seen from the dining room, and not infrequently at the top of the steps one finds a communal terrace, behind which today's architect has placed the actual living rooms of the family, directly on the open lawn: house and garden are always intrinsically connected, their parts complementing and defining each other with increasing intensity. In this manner, these accomplish nothing other than the business of type formation.

Exactly the same process in recent *Landhaus* building today has already led to universally valid norms and even further subdivisions, and indeed in this sense our most important architects build "typically." Only petty show-offs and pseudo-individualists "never repeat themselves." And that such developments don't lead to poor end results for the life rhythms of the top ten thousand is shown by England. A journey through England, especially through its landscape, has such a pleasant effect on us, not least because everywhere we discover a kind of willing uniformity of country seats, houses, and gardens, resting on the firm ground of a common culture based on the typification of property.

Thus, I don't hesitate to declare how I myself imagine *whole gardens as types* in the not too distant future. It is self-evident that, in addition to other influences, the living nature of garden materials places limitations on this. It is also self-evident that this development will first take place in the gardens of the hundred thousand, the small and smallest workers' and garden-city gardens. *But they will come.* The more the well-to-do classes come together in professional circles on the one hand, and the more they distinguish themselves from each other socially on the other, then the more appropriate a type-garden becomes for them, when also conceived in more flexible form. The type, that is the true reality! Following on that, I ask that the examples published here once again be individually examined in relation to the typical forms within their organism.

For additional clarification of this discussion, I have selected from my own practice a garden for a house designed by Professor Riemerschmid.[48] Here the circumstances were advantageous. The correct positioning of the house in relation to the sun, the site, and the levels. Clear, unambiguous, honest division, and generous proportioning of the spaces. Separation of professional practice from family life, from living and service, an exact knowledge of each family member, of their personal tastes and their social identities—all of these carefully balanced and adapted to the situation. A garden resulted from this, of which I must say, if I were presented with this problem again, I would have to solve it in *exactly the same way*: it virtually became the *garden type for a physician.*

Garden of Dr. Naumann in Riesa [near Dresden]
L. Migge

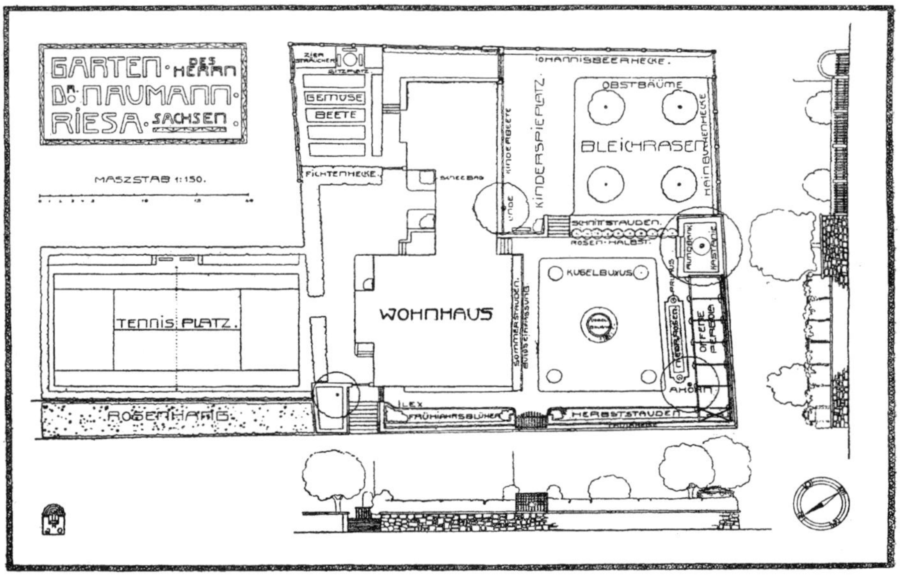

Plan of the above garden
L. Migge

D. Garden Art Nevertheless?

Is this a lack of imagination? No, it is a restraining tendency, a self-reflection, and a limitation to the essential. It is called *organizing*.

But indeed it is also—art. Thus, a perfect type, in whatever field that may be, always holds within it a spirit of creation, a spark of divine power, and a true artwork is wholly inconceivable without typical characteristics, indeed, these are its formal legitimation. In this connection, let us again look at history, for references such as Homer and Phidias, Dante and Kant, even Jesus Christ, Buddha, and Mohammed, who on the deepest level were nothing other than powerful figures moving the world toward typical forms, ideas, and feelings. And how much more valid is this, when we descend from this sublime region to the more modest level occupied by the anointed of our time. Here we see typifying work even more clearly. Some among our contemporary pioneers are destined to try to show us all the paths, even when these are still obscured by mist, to art, to *our art*.

Admittedly, this is the work of singular individuals and the great. But the rest of us are not less, and our work is not irrelevant, even when it appears that we have less to achieve. We are the foundation. The rhythm of those harmonies that naturally arise from our thousandfold works, which until now has been almost ignored as a side product, *that* is the real fuel for potential forms, which our species could be destined to recognize and realize under its leader. This is also true for the garden. Academic reflections and aestheticized feelings are not capable of formulating our own garden, as much as they would like to be able to directly contribute. *Instead, this will be brought about through the rhythm of work, which arises out of the varying, constant utilization of all small spiritual impulses and real actions that together constitute stimulating garden living:* this is the true rhythmic basis of the garden culture of the twentieth century! A garden culture of nature and art in one. Here lies the next, and indeed also the only way, to happily renew the world of the garden.

Following this, it is *first* a matter of developing, defining, and propagating such an intensive form of garden living, and then in the *second* place, of establishing specific concepts of form and aesthetic directions. From my perspective, too many formal intentions tend to diminish rather than enhance rhythmic creations. *A universal and equal garden will is essential!* Though artistic qualities may be attributed to our applied gardening, it thrives most naturally through the most straightforward combination of all of the lower and higher purposes in question here. This suffices in all cases, but only when a degree of spontaneity, repose, and generosity prevails, along with a minimum of taste. This suffices particularly for the prevailing majority. The extraordinary tasks that come with a higher sense of responsibility today will also not suffer for lack of those with refined nerves and talent.

So it indeed appears to follow as observed from the rhythmic side as well, not only because more promising and therefore more desirable, but out of direct necessity, namely: to have *many gardens*. Gardens in great numbers, gardens for everyone, and for every purpose, will not only outwardly result in more peace and happiness among people, rather, this multitude of gardens itself will be the cause of it all, which in fact was first made possible by our new garden rhythm. Just as Leonardo da Vinci would not have been the creator of the *Mona Lisa* (and this certainly could not have been stolen with such pleasurable excitement a half-millennium later),[49] if hundreds of Madonna's had not been painted in the context of a feverish Madonna cult *before* Leonardo—and such breathtaking garden images, such charming and unique vegetative refinements as found at the Villa d'Este in Tivoli for example, would also be inconceivable without precedents.

Then onward. There is indeed nothing else left—so let us do it purposefully and well: *let us found a new, joyful cult of the garden!*

That was "garden art." Why do I speak of it anyway, when I really want to discredit it? Because I want our garden makers to finally rid themselves of this *talk* about things, that in fact need to be *worked on*. One sees how dangerous this is overall: in my effort to open up a free path toward a well-intended defense against the overblown garden aesthetic in our lives today, I virtually have become—horror of horrors!—a silver-tongued proselytizer myself.

THE GERMAN GARDEN

VIII

A. Social Gardens

It must be ever more emphatically declared: for us it cannot just be a matter of having gardens that look good and are harmonious; instead the question is: *How can the garden bring happiness to contemporary people*? We already know that there were times when this obligation was more or less fulfilled. And even if this was for certain groups and the privileged classes, that is, *segments* of the population who provided a pleasant background to the garden, nevertheless its influence on the lives of the people as a whole was then still so strong, and its effects on following generations still so great, that these few gardens from history are as alive to us as if we could see them today.

And from all of this we note: our future garden will not, and *must* not, be the preserve of the privileged among the wealthy or intellectuals. Our garden culture, if it is to come about at all, is only *conceivable on an immeasurably expanded basis in contrast to what has come before.* And then in its realization how much greater the effect will be on the elevation of humanity!

Today we are a people of sixty-five million; in twenty years that may be a full eighty. Then how many will enjoy the pleasure of a garden? Only a few. In Germany in the year 1907, one counted eight hectares of actual gardens in all categories, per one thousand citizens. This means that when one calculates a third of all vegetable gardens and orchards as purely commercial facilities, and a further third as the many useless or only partially useful front gardens, institutional gardens, club gardens, and public areas, then for every one thousand people there is slightly more than two and one-half hectares of private gardens. When each of these gardens takes up only a Morgen (2,500 square meters) on average—and there are many parks that boast more than ten times that—then among one thousand Germans only ten have their own gardens. Not all of the well-to-do among us own a garden, and since rural districts are included in this calculation, then finally, at most each group of one hundred families in the metropolis has even one. *One hundred German families must share one relatively large garden.* In terms of health and morals, i.e., the national economy in the highest sense, this is a truly shocking figure, particularly when one realizes that for every garden that is lacking, the counter-factors of alcohol, sickness, and an unhappy family life can be added into the calculation. In truth, this leaves no room for sentimentalities!

In spite of this, we will progress as a people. All the classes will progress. We live and work and become prosperous. Wealth comes to our country, and then follows—heaven willing—a little *time* and *leisure*. That is indeed the correct foundation on which a great collective garden culture can come about. Making sure that this happens is the task of our generation. *It is our obligation and at the same time our clear advantage, when we explore what is special about our work and our living, the effects and potentials of our economic life, and our entire intellectual sphere, for our sprouting gardens.* To consciously and tirelessly use that will, these inalterable processes, which are indeed the forces driving our contemporary existence, as well as the goals of a heightened garden living—that is the will to garden culture. But in order to experience it, above all we must first have *the* garden for this. Ownership of a garden in this era must depend upon nothing other than the *desire* to have one: *a garden for everyone*!

The new German garden is either a social concern, or seriously none at all.

B. The National Economy and Gardening

This garden, in the best sense a social garden, is in fact the future garden of the world everywhere. Then the conditions that bring forth new garden living among us today are the same for all other peoples, or they will be in the foreseeable future. They too are already surrounded by the spirit of the twentieth century, which embraces science, technology, and transportation; they too are under the yoke of capitalistically organized labor, by which the masses concentrate and expand. But gardens automatically bring this about as well: *Europe and the whole world need gardens*. Then a certain aspect of the world economy immediately comes into the debate. The new German gardening goes beyond the intellectual-ethical, and is even itself a powerful factor in the material and purely economic sense. Or it should and in fact will be.

In principle, there was much about the national economy that could have been read between the lines up to this point. We spoke about this rather plainly because it had to be shown that what is particular about the way we work and live in the big cities literally demands gardens, both defining the other. It was the case when we spoke of urban planning and the house and dwelling unit, that each of these

plate 22

Gray willows, reeds, and Nymphaea unite to form a lush waterside image. In the background, the water pavilion (a new type) is reflected.

PLATE 22

categorically needs green space and flowers although each in its way. National economic problems were touched upon in the chapter[50] on the people and personalities who influenced, nurtured, and indeed produced the more intensive horticulture—affected in so far as experts and laypeople more or less consciously continued to generate gardening ideas. In the final, most important sense, when we occupy ourselves with plants and plant growing, or with wood, iron, and stone in the garden, this concerns the national economy. In short, whenever we were talking about the intellectual and material production of gardens and garden elements we were also always talking about the economy at the same time.

The national economy in fact signifies nothing more or less than to *produce* and to *utilize*. Whether a matter of steel, paintings, or gardens, it is all the same when conceived in terms of economics. Every actual achievement has an economic impact. Thus, we immediately produce gardens and garden meaning when work, wealth, and hygiene serve universal inward and outward needs to this end. But in order to contribute to the national economy through the garden in a *conscious* and *methodical* manner, it is necessary to further examine and organize the means, experience, and ideas gained in this manner, to develop a wider, more certain field of application: the new German gardening must first be economically organized.

If one will, in this sense the garden cultures of earlier times also represented the national economy, even when that has not always been very apparent. Thus, in principle, when Syrian, Moorish, or medieval gardeners traveled through the world of the cultivated and wealthy at the time for the purpose of selling their gardening knowledge, they in fact were engaged in an economic act retroactively affecting all of the people. Or when Le Nôtre with his staff planned luxurious gardens for the princes of the world, drawing upon countless laborers and materials from within France. In spite of this, the garden culture of the technical era, when it arrives, will be something completely different from that of the former Orientals or French, again in respect to the national economy. In former times, gardens of a more elevated nature were in fact few and far between, so that despite all the high achievements in gardening, these simply could be nothing more than unique specimens. By contrast today, a hundred or a thousand times more gardens are laid out. Everyone needs gardens. We need *gardens in mass numbers*. This indeed gives our movement an infinitely heightened cachet in terms of the national economy.

Produce and utilize. Naturally it is truly a great thing, to discipline a people like ours in every aspect related to the garden, so that we may be able to speak of our homeland as the country of gardens. It is great, because in the sense of the national economy alone, this internal process must call forth an entirely new *way*

plate 23

Is there anything more pleasant than admiring the light-dappled splendor of the garden from a pretty vine-covered seat?

PLATE 23

of working [*Arbeitsleben*]. Many new gardens will need new heads, hands, and new materials. New market gardens and nurseries will be established, and the old ones will be reorganized and expanded. Capital, which has thus far mostly avoided the field of gardening, will become more interested and engaged. As a result, new businesses and industries for the array of garden embellishments (ceramics, artificial stone, woodwork, etc.) that are just now emerging will become established, and many more besides. An economic wave of undeniable strength and duration will go hand in hand with the propagation of this gardening will in Germany.

But as important as this is, it doesn't suffice as the goal. It is insufficient, because these economic and intellectual efforts must also be *simultaneously* directed outward in order to foster and deepen this domestic process, since it is essentially the aggressive force and mental intensity of the *first stage of development* that make an impression abroad. No intellectual or economic movement of any significance can exist for long in isolation from its environment without doing damage to itself. This must also be the case with modern gardening: now or never! Once people abroad have learned from us, then they won't need us anymore. It is so: as a gardening people we would be finished the moment that those abroad begin to ignore our garden work here at home, even if it is shown to be superior. For all of these reasons, a renewal of the German garden that is driven by the national economy must be directed toward countries abroad from the beginning. *We are about to create a garden culture, with singular examples characteristically expressing the unity of a people and the future within it, to become a sign of a cosmopolitan whole.*

As outlined here, this great obligation excludes from the beginning the possibility that any one individual alone could meet it. More significantly, here a *national* task of the most far-reaching importance lies open. National intellectual and economic battles are the most important factors today determining whether a nation is to be or not. Just as today, when the powers of peaceful penetration gradually overcome those of fading aggression within the fiber of the people, in future the people who are the very first to bring the domestic evolution of all spheres of culture to full realization, will be the ones who are the most superior, most successful, and most happy. Now in the era of the global exchange of all goods, an achievement can only be considered to be of national significance when it has the capacity to have an international effect. In this sense, the new gardening belongs to one of the greatest national endeavors that I can imagine.

Thus, it will essentially be a matter of trying to spiritually and materially imprint a certain cosmic tendency upon our young garden movement, as we consciously experience it. However, since other nations that are more blessed than we in terms of gardening think and work in similar ways—only the Philistine believes that others sleep—then even the external form of this aspect of national development will again be a battle. Indeed, even the finest goods can and must be fought over with the greatest energy. Then here is truly new territory: here the capable will soon win in a field of civilization hitherto undefined. Here there will be conquerors and conquered. But Germany doubtlessly holds the best position. And I believe that something more than merely a desire to be self-sufficient will incite us to take in hand this important, promising area within the world economy. New gardens: "made in Germany!"[51]

And thus after being calm and reserved for so long as I believed necessary, my book culminates in this chapter with a call to arms. Toward the cheerful and joyful battle for the supremacy of the garden! Rushing forward toward the transformation of our own land into a land of flowering gardens in which everyone, even the poorest, have their share. And also forward, where it means securing and improving the working conditions and living standards of all those involved with gardens, through an extensive, peaceful invasion of foreign countries. We want to be the people of the *many and delightful gardens*—that is the new and meaningful sense of a garden culture of the twentieth century.

Notes to the Translation

1 "Garden-web" is an approximate translation of *Gartenweben*. The verb *Weben* means "weaving" or "to weave," but it is also commonly used in a figurative sense to mean "busyness." Migge was making a play on words, associating the figurative use of this verb with "veil" later in the sentence.

2 The page numbers in brackets refer to the translated edition.

3 The numbers 3 and 7 have clearly been switched in the original reference here.

4 Pliny the Younger, first century AD.

5 Probably a reference to the German Peasants' War (*Deutscher Bauernkrieg*, 1524–26), which was one of the most important revolutions in Europe. Migge is suggesting that after the revolt the farmers were less dependent on the nobility, allowing them the relative freedom and material capacity to create gardens.

6 Members of the Fugger family were immensely wealthy bankers in the fifteenth and sixteenth centuries. In the early sixteenth century, the family donated money to establish a complex of almshouses in Augsburg, known as the "Fuggerei." In Migge's era, the Fuggerei was thought to be an advanced form of housing design and was often mentioned in German garden-city literature.

7 Hans Sachs (1494–1576) was a "*Meistersinger*" and poet in Nuremberg.

8 "Bitter oranges," *Pomeranzen*.

9 The *Städtebau* exhibition held in Düsseldorf in 1910 was curated by Werner Hegemann; it was for this exhibition that Hegemann succeeded in convincing Daniel Burnham to send the paintings by Jules Guérin of the Plan of Chicago, exhibited there for the first time in Europe (see Introduction, pp. 23 and 37n7).

10 The Schrebergarten was named for Daniel Gottlob Moritz Schreber (1808–61), a German physician in Leipzig who argued for the need for outdoor exercise for children in urban areas. In 1864, Leipzig school principal Ernst Innozenz Hauschild established the first Schrebergarten by leasing land for the physical exercise of children. The term Schrebergarten is now a generic expression for allotment gardens all over Germany.

11 "Garden pavilion" is an approximate translation of *Gartenlaube*, which can mean "garden house," in the sense of a small pavilion; there was also a popular journal of this name, *Die Gartenlaube*. *Laubenkolonien* is thus translated as "garden-pavilion colonies."

12 "Bourgeois" as a translation for "Bürger" is not pejorative in this case but rather means something like "upstanding middle-class citizen."

13 "Dairy camp" in this case refers to facilities where "weak children" observed milk cows and consumed their milk; milk was considered a healthy drink for all, and was provided in dairy bars for adults as an alternative to alcohol, sometimes in public parks.

14 The term *Landhaus* is used here, rather than the English "country house," because the German term has specific connotations, as explained in the Introduction. The *Landhaus*, as popularized by the German architect Hermann Muthesius, was more a suburban villa than an actual country house.

15 This phrase is attributed to New York City housing reformer Jacob Riis (1849–1919), although it has been variously paraphrased by other authors. Migge probably took the German translation from Werner Hegemann.

16 "Commended to the care of the public" translates *dem Schutze des Publikums empfohlen*, now an outdated expression, but in Migge's time common on plaques in public places, such as on statues.

17 The Summer Olympics were held in Stockholm in 1912.

18 A popular swimming lake in southwest Berlin.

19 "Columbia" refers to the District of Columbia—that is, Washington, D.C.

20 "Barbarossa," literally meaning "red beard" in Italian, was the epithet of Holy Roman Emperor Friedrich I (1122–90).

21 Emanuel von Seidl (1856–1919), a successful architect and designer in Munich.

22 The equivalent German phrase for "in the country" literally means "on the land," hence the play on words here.

23 In this rather fanciful construction, Migge implies that Adam was emotionally moved by authentic nature "before" the fall from Grace—that is, before he tasted the apple in the Garden of Eden—and afterward he convinced himself that he could still have the truly natural garden, although it had thus been lost forever. In this sense, Migge is comparing Adam to those who still believed in the "natural" garden, which Migge regarded as a nonsensical impossibility.

24 An important German garden city, Hellerau was founded close to Dresden in 1909 with close connections to the Werkbund; many of its buildings were designed by Hermann Muthesius. See, for example, Michael Fasshauer, *Das Phänomen Hellerau* (Dresden: Hellerau-Verlag, 1997).

25 Peter Behrens (1868–1940), painter and architect, active in the Werkbund and in garden reform, extremely important for German modernism; Max Laeuger, see the translator's note, p. 95; Joseph Maria Olbrich (1867–1908), one of the founders of the Vienna Secession movement, who conceived of the single-color garden (*Farbengarten*) alluded to by Migge (GK 110, GC 151); Hermann Muthesius, architect, see notes in Introduction; Friedrich Bauer (1872–1937), garden architect; Reinhold Hoemann (1870–1961), garden architect based in Düsseldorf, editor of the professional journal *Die Gartenkunst* from 1911 to 1913.

26 The phrase "special materials" seems to mean vegetal matter, which is "dependent" on factors such as the weather, in distinction to less mutable architectural materials.

27 From the prologue to the first play in Friedrich Schiller's dramatic trilogy *Wallenstein*: "von der Parteien Hass und Gunst verwirrt." Friedrich Schiller, *The Camp of Wallenstein*, trans. James Churchill, in *The Works of Frederick Schiller*, vol. 2 (London: Bonn, 1846).

28 François-Adrien Boieldieu composed a comic opera called *La Dame Blanche* (*The White Lady*) based on five of Walter Scott's novels, first performed in 1825.

29 A "reform dress" would be loose and flowing; during this period, reformers advocated more freedom for the human body, particularly by abandoning the corset.

30 The important commercial nursery Haage & Schmidt was established on the eastern periphery of Erfurt in 1863; see Ilsabe Schalldach, "Die Familie Haage: 325 Jahre Gärtnereigeschichte," in *Blumenstadt Erfurt: Waid—Gartenbau—iga/egapark*, ed. Martin Baumann and Steffen Raßloff (Erfurt: Sutton Verlag, 2011), 74–107.

31 See, for example, Hans Leopold Zollner, "Nicolas de Pigage: Sein Leben und Werk," *Badische Heimat* 4 (1975): 115–27.

32 "'The Temple Show' was officially the Great Spring Show of the Royal Horticultural Society. The Great Spring Show was first held at Kensington in 1862, and when the

Kensington garden was abandoned, it was transferred to the Temple Gardens and held there annually until 1911. In 1912 the show was cancelled in order to make way for the Royal International Horticultural Exhibition, held in the grounds of the Chelsea Hospital. The following year the Great Spring Show was moved there, and has been held there ever since." Brent Elliott, Historian, RHS Lindley Library, London, e-mail message to author, April 2012.

33 Here Migge is probably referring to the work of the biologist Ernst Haeckel, who described nature as being like an artist; his drawings of deep-sea creatures were partly intended to challenge traditional concepts of beauty, emphasizing inner structure and function in relation to outer form.

34 A shrub or sapling with multi-stems at its base.

35 Gaius Cilnius Maecenas (70–8 BCE) was an important patron of the arts and a friend of Octavian (who became the first Roman emperor as Caesar Augustus).

36 Max Läuger was a ceramicist, painter, and designer of gardens and houses. Migge made his acquaintance because of his importance as a garden designer. The surname can be spelled with or without an umlaut over the "a." See Rudolf Velhagen and Maximilian Ulrich Schumann, eds., *Max Laeugers Arkadien—Keramik, Garten, Bau, Kunst* (Baden: Museum Langmatt, 2007). An important garden-design and horticultural show was held in Mannheim in 1907, which signaled an important victory for the cause of German garden reform.

37 "Nuggets," *Brocken*; the German term may refer to "clumps of earth," which I have translated in the sense of "nuggets of knowledge."

38 The "Mark" refers to Brandenburg and its sandy soil; "lower Germany" is the area including Lower Saxony (Niedersachsen).

39 Possibly a reference to Genesis 1:11–21: "Und Gott sprach: Die Erde lasse Gras hervorsprossen, Kraut, das Samen hervorbringt. . . ."

40 "Pick" is a conjectural translation of *Bigge*, a rare word with uncertain meaning. In old Lower Saxon (Niedersachsen) dialect, *Bigge* refers to a clump of earth.

41 Gert Gröning has speculated that this passage refers to Migge's garden design in Roggendorf, shown here in Plate 21: Gert Gröning, "Die 'Alster' von Roggendorf—Leberecht Migge und der Park von Haase: Eine Betrachtung aus hundertjähriger Distanz," in *Bildende Kunst in Mecklenburg und Pommern von 1880 bis 1950*, ed. Bernfried Lichtnau (Berlin: Lukas, 2011), 497–517.

42 From Goethe's poem "Kunst und Natur": "In der Beschränkung zeigt sich erst der Meister" (Migge omitted the words "zeigt sich"). Translation from Angela Esterhammer, *Romantic Poetry* (Philadelphia: John Benjamins, 2002), 189.

43 The "Biedermeier era" refers to a period in German-speaking countries between the end of the Napoleonic Wars, in 1815, and the revolutions of 1848; this was a period much celebrated by design reformers in the early twentieth century, as it was characterized by restrained classicism as opposed to the more ornate historicism of the late nineteenth century.

44 "Ruberoid" is a brand of roofing material still produced in the United Kingdom.

45 Certain of Runge's paintings were celebrated by garden reformers as being examples of "good" Biedermeier-era garden design.

46 Here Migge makes a play on words between "life" and "experience" (*Leben* and *Erleben*), which is not really translatable into English. Migge also italicized the prefix *er-* for further emphasis, as I have done in brackets.

47 The poet Richard Dehmel (1863–1920) lived in the Hamburg suburb of Blankenese, not very far from Migge.

48 Richard Riemerschmid (1868–1957), an architect, planner, and painter from Munich, was a founding member of the Werkbund.

49 The *Mona Lisa* was stolen from the Louvre in 1911; it was hidden by the thief for two years, then discovered in Florence and exhibited around Italy, and finally returned in 1913.

50 "Chapter" (*Kapitel*) is misprinted in the original as *Kapital,* or "capital," which makes no sense in this context (GK 154).

51 The phrase "made in Germany" is in English in the original.

TABLES

TABLE 1
Playgrounds and sport areas

City	Public play, sport, and exercise areas				Private playgrounds		Public and private playgrounds		
	Number of	Areas in hectares	Construction cost	Maintenance costs	Number of	Areas in hectares	Exercise areas in hectares (1907)	Exercise areas in hectares (1911)	Permanent playgrounds in 1910 (in hectares)
1	2	3	4	5	6	7	8	9	10
1. Cities with more than 200,000 inhabitants									
Berlin	—	—	—	—	—	—	36.00	—	10.00
Breslau [Wrocław]	84	86.86	10,585	18,350	—	—	21.38	26.86	90.00
Chemnitz	10	11.36	110,000	9,000	—	—	6.71	11.36	10.00
Cologne	55	51.90	—	—	—	—	2.20	51.90	18.00
Dortmund	9	7.30	4,500	1,500	5	5.70	2.00	13.00	—
Duisburg	2	2.70	16,450	300	6	5.10	15.0	7.80	9.00
Hamburg	46	8.72	—	36,420	21	38.20	40.00	46.92	10.00
Hanover	19	70.44	—	—	43	42.00	15.00	**112.44**[1]	—
Leipzig	—	56.00	—	688	—	—	35.79	56.00	—
Magdeburg	4	4.50	—	1,068	—	—	8.84	4.50	13.00
Stettin [Szczecin]	24	3.17	—	—	4	—	5.50	3.17	2.00
Stuttgart	177	53.10	—	—	30	31.00	15.30	84.10	20.00

TABLE 1

(continued)

City	Public play, sport, and exercise areas				Private playgrounds		Public and private playgrounds		
	Number of	Areas in hectares	Construction cost	Maintenance costs	Number of	Areas in hectares	Exercise areas in hectares (1907)	Exercise areas in hectares (1911)	Permanent playgrounds in 1910 (in hectares)
1	2	3	4	5	6	7	8	9	10
2. Cities with 100,000–200,000 inhabitants									
Aachen	10	9.00	—	2,000	3	1.75	8.00	10.75	5.00
Kassel	6	0.47	—	420	—	—	—	**0.47**[2]	—
Danzig [Gdańsk]	6	9.20	41,600	3,500	1	0.60	2.50	9.80	6.00
Elberfeld	7	5.36	—	3,000	1	0.75	—	6.11	—
Halle	1	8.1	550	—	12	5.91	6.00	14.01	12.00
Mannheim	3	2.10	4,500	650	4	1.76	44.41	3.86	72.00
Mühlheim on Rhine	4	4.90	60,000	7,612	6	3.50	2.73	8.40	—
Wiesbaden	1	1.40	—	1,000	—	—	1.37	1.40	2.00
3. Cities with 50,000–100,000 inhabitants									
Bonn	12	10.08	—	5,600	6	4.42	22.40	14.50	—
Frankfurt Oder	4	0.80	—	—	4	0.90	0.80	1.70	—
Hagen	5	0.80	5,650	1,100	4	3.13	4.50	3.93	—
Ludwigshafen	9	5.30	13,000	3,330	8	8.90	6.64	14.20	—
Osnabrück	1	3.59	—	500	2	3.00	42.07	6.59	—
Ulm	3	1.50	—	500	6	5.50	1.58	7.00	—

1 This outstanding number of playgrounds for Germany reflects Hanover's reputation in sports.

2 It is typical for the understanding of the purpose of public facilities in Germany up to now, that even cities with unusually high amounts of green areas possess disproportionately few playgrounds; thus, for example, Kassel with 2,150 hectares of public facilities has only **0.47 hectares** of playgrounds.

TABLE 2

Overview of public squares in English cities with more than 200,000 inhabitants in 1902[1]

City	Population	Number of objects	Hectares	Purchase cost marks	Square meters per capita
London (County)	4,536,000	313	2,412.80	43,980,000	5.30
Manchester	543,000	37	395.70	2,564,000	7.20
Leeds	429,000	20	298.40	6,124,000	7.00
Dublin	373,100	14	574.00	72,000	15.40
Belfast	349,000	7	104.40	210,000	**3.00**
Bristol	329,000	21	270.80	1,764,000	8.20
Edinburgh	316,500	24	518.40	2,834,000	**16.30**
Total	6,875,500	436	4,574.50	57,548,000	8.90

1 These figures are inclusive of all accessible green spaces used by the public.

TABLE 3

Overview of park, garden, and ornamental areas in German cities with more than 200,000 inhabitants[1]

City	Year	Inhabitants	Overall [urban] areas in hectares	Public woods in hectares	Public parks in hectares	Public parks as a percentage of overall [urban] areas in hectares	Square meters per capita
Berlin	1906	2,070,695	6,352.25	—	457.60	7.15[2]	2.20
Breslau [Wrocław]	1910	511,891	4,228.77	2200.00	264.20	6.30	5.10
Charlottenburg	1906	239,559	2,150.60	—	89.10	4.20	3.70
Chemnitz	1906	244,927	3,971.41	—	153.20	3.80	6.20
Cologne	1911	516,167	11,133.31	76.00	257.40	2.30	4.90
Dortmund	1910	214,333	3,073.60	88.30	28.40	0.90	1.30
Dresden	1907	546,882	6,755.60	117.60	253.60	3.70	4.60
Düsseldorf	1906	253,274	4,869.04	—	131.20	2.70	5.20
Duisburg	1910	229,478	7,072.70	447.90	38.70	0.50	1.70
Essen	1907	294,629	2,776.84	—	23.00	0.80	0.70
Frankfurt Main	1910	414,598	9,390.66	34.7	124.40	1.30	3.00
Hamburg	1911	1,007,000	12,354.81[1]	—	162.13	1.30	1.61
Hanover	1906	250,024	3,956.89	—	186.20	4.60	7.40
Kiel	1911	211,044	2,371.55	117.00	53.60	2.20	2.50
Leipzig	1908	587,635	5,696.82	803.60	110.00	1.90	1.80
Magdeburg	1906	240,633	5,550.04	—	298.50	5.30	**12.40**
Neukölln	1910	237,378	1,187.58	—	7.70	0.60	**0.30**
Nuremberg	1907	332,651	6,419.52	9.80	116.90	1.80	3.50
Stettin [Szczecin]	1910	236,145	6,722.43	299.20	47.60	0.70	2.00
Total		8,698,943	—	—	2,803.43	—	3.69

TABLE 3

(continued)

City	Year	Inhabitants	Overall [urban] areas in hectares	Public woods in hectares	Public parks in hectares	Public parks as a percentage of overall [urban] areas in hectares	Square meters per capita
Cities with 100,000–200,000 inhabitants							
Aachen	1910	156,044	5,061.83	1,279.00	62.20	1.20	3.90
Augsburg	1907	94,923	2,186.31	359.10	46.90	2.10	4.90
Bochum	1906	118,464	2,064.39	—	30.10	1.40	2.50
Braunschweig	1907	136,397	2,767.00	—	69.00	2.50	5.00
Danzig [Gdańsk]	1910	170,347	3,667.61	1,411.70	99.00	2.70	5.80
Wilmersdorf	1910	109,729	836.18	—	10.00	1.20	0.90
Karlsruhe	1910	134,185	3,241.64	780.40	46.30	1.40	3.40
Kassel	1900	106,000	2,152.16	—	213.00	9.70	**17.70**
Mülheim on Rhein	1910	112,602	5,666.00	1,688.70	0.20	0.03[3]	**0.02**
Posen	1906	136,808	3,303.70	—	51.10	1.60	3.70
Schöneberg	1906	147,388	946.38	—	8.40	0.90	0.50
Wiesbaden	1911	109,033	3,606.84	—	47.80	1.30	4.40
Total		1,531,920	—	—	684.00	2.10	4.30

1 In contrast to those of the English cities, these figures, with few exceptions, stand for decorative ornamental areas, with only the smallest proportion for parks that are entirely usable.

2 This unusually high percentage of park areas in relation to overall [urban] areas is explained by the extremely dense occupation of the city of Berlin (due to the presence of tenement houses).

3 Representative of the lack of park policy and planning among rapidly growing industrial cities.

TABLE 4

Municipal allotment gardens in German cities with more than 100,000 inhabitants[1]

City	Unbuilt municipally owned areas in hectares	Municipal allotment gardens in hectares	Municipal allotment gardens		Expenditure for municipal gardens, specifically for:		Annual municipal income for leases	Length of lease	Lease rate per square meter in pennies	Playgrounds in allotment colonies in square meters
			Number of	Area in square meters	Initial construction	Annual maintenance				
Berlin	317.00	15.00	500	200	—	—	—	6 years	3.5	3,000
Bremen[2]	—	45.91	550	250–3,000	—	—	—	6 years	2–5[2]	—
Kassel	686.00	0.60	**18**	260–340	—	—	ca. 270	indefinite	5	—
Cologne	147.78	—	253	150–530	38,000	2,260	8,200	indefinite	7–9	1,500
Dresden	377.25	—	24	200	donation	760	2,000	indefinite	**30**	250
Elberfeld	350.41	0.40	22	150	800	—	132	indefinite	4	—
Erfurt	318.31	30.60	523	500	—	—	7,058	12 years	6	—
Frankfurt Main	4,576.80	43.30	950	100–1,000	—	500	10,504	6 years	2–8	2,000
Halle on Salle	497.70	1.86	111	100–200	9,000	50	2,710	1 year	15	—
Hanover	1,492.90	60.00	1,200	300–2,500	—	—	16,500	6 years	1–8	—
Kiel[3]	789.50	209.00	**4,370**	420	170,000	3,000	79,820	indefinite	4–6	—
Magdeburg	992.90	19.20	386	ca. 360	46,562	2,075	12,911	6 years	8	5,225
Mannheim	2,306.80	22.90	869	160–350	119,725	8,980	13,365	10 years	4–7	—
Neukölln	?	28.00	1,600	?	1,000	—	9,000	1 year	4–5	?
Wiesbaden	181.77	5.00	141	100–200	—	1,200–1,500	3,322	indefinite	10–16	—
Municipal allotment gardens in German cities with 40,000–100,000 inhabitants										
Dessau	—	7.70	360	180–230	25,000	1,550	2,363	6 years	6–8	4,680
Elbing	191.29	1.20	52	150–240	4,000	—	800	3 years	10	1,000
Flensburg	417	—	500	200–1,000	—	400	—	1 year	3–4	600
Frankfurt Oder	1,159	1.70	32	200	50	20	239	1 year	4	1,000
Halberstadt	—	1.90	500	200–400	5,000	100	830	6 years	5–8	3,500
Kattowitz [Katowice]	—	2.20	30	100–400	—	—	—	—	15	2,000
Lübeck	—	25.00	700	300	—	—	—	10 years	2–3	4,000
Mühlheim on Rhein	—	2.40	58	250–320	8,054	200	1,030	indefinite	6	1,725

1 Of note almost everywhere are the short and unconditional lease termination notice periods, as well as the difference in lease rates.

2 General leaseholders.

3 The initiative of Kiel is to be lauded, which along with Leipzig has broken new ground in the important area of communal garden policy (Kassel as counterexample, see table).

APPENDICES

appendix 1

One of the most positive products of new urban park policies: THE INNER PARK. It is reserved for the use of the occupants of the surrounding building blocks, which also contain additional green spaces within them.

Wolf-Schöneberg

TRANSLATOR'S NOTE

Paul Wolf (misspelled in the original as "Wolff"; 1879–1957) was an architect and urban planner. Illustrated here is the project for the Ceciliengärten in Berlin, which was completed after World War I. See Eva Benz-Rababah, "Leben und Werk des Städtebauers Paul Wolf (1879–1957) unter besonderer Berücksichtigung seiner 1914–22 entstandenen Siedlungsentwürfe für Hannover" (PhD diss., Hanover University, 1993).

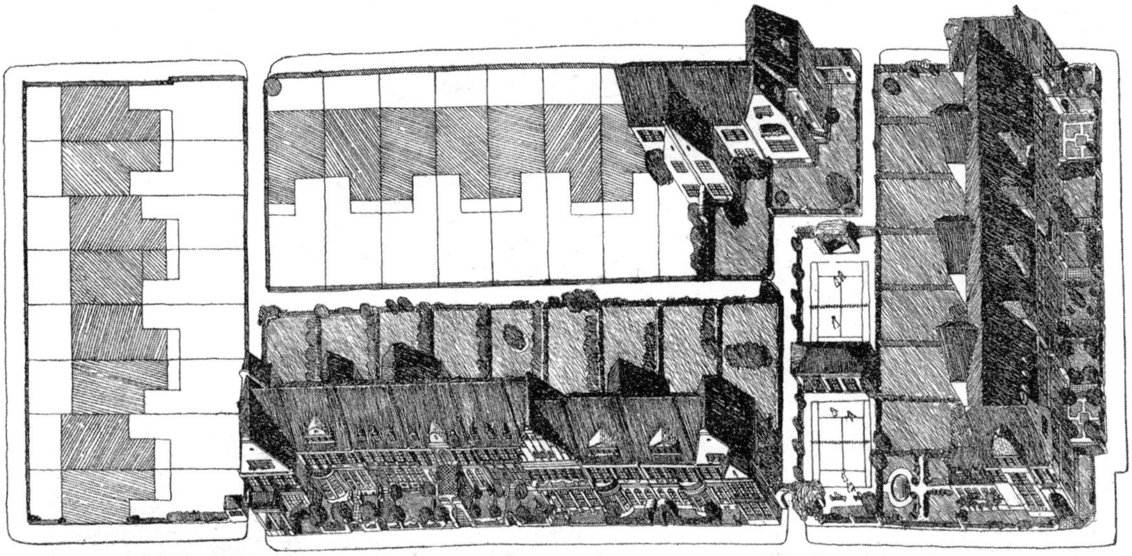

appendix 2

A STAGGERED METHOD OF BUILDING, which allows the densely populated [apartment] houses on the edges of the block to have shared play areas and vegetable gardens, while the lower (single[-family]) houses on the interior have their own individual gardens.
Wolf–Schöneberg

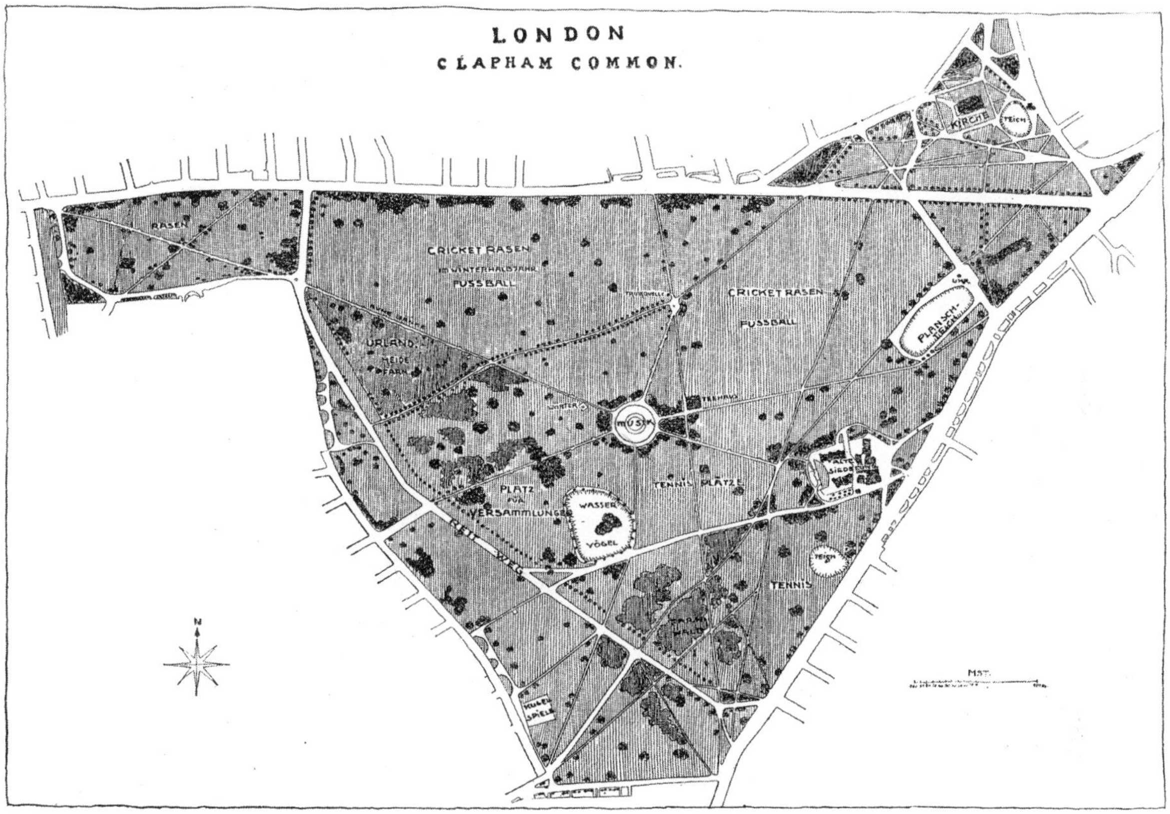

appendix 3

A TYPICAL ENGLISH PARK. Extensive, grove-like, accessible meadows, traversed by casual paths. Water, as it was found. (Even an old settlement was not an impediment.) A music pavilion with a teahouse next to it, and a simple enclosure around it: the whole not much more than a naively utilized but indeed blessed landscape.

London County Council

TRANSLATOR'S NOTE

Illustrated is Clapham Common in southwest London; Migge probably saw Clapham Common when in England on the study tour of the German Garden City Association in 1910; extant.

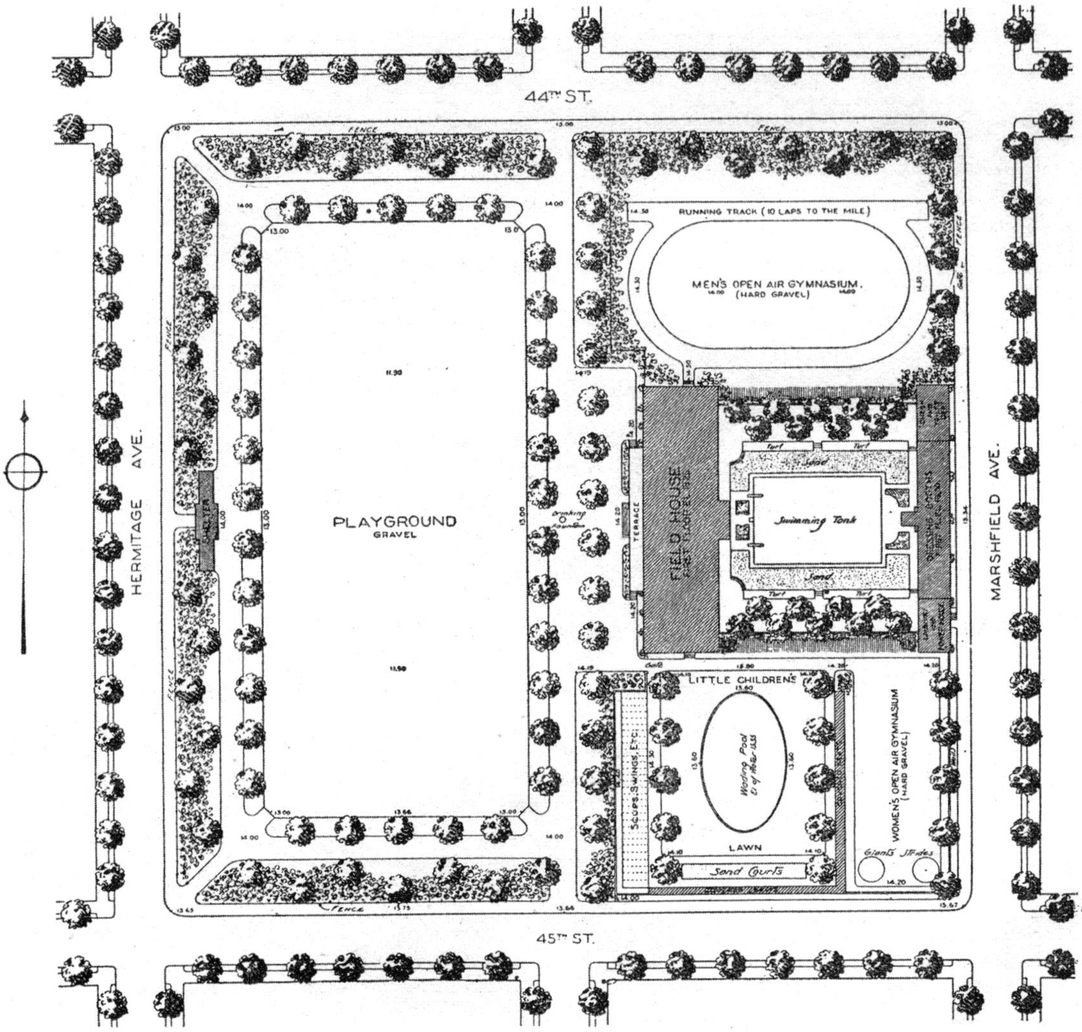

appendix 4

A REAL PLAY PARK TYPE (Nr. 14), the City of Chicago. Every square meter is rationally utilized, with playing fields, gymnastic areas, water oaks, sunbathing and swimming areas, for both sexes. In the middle is a "field house," of about seventy meters in length, with dressing rooms, community rooms, and a library. (See also Plate 5, lower.)

Olmsted Brothers, Landscape Architects (Brookline, Mass.), 1904

TRANSLATOR'S NOTE

Illustrated is Davis Square Park in the "Back of the Yards" area of Chicago, near the former stockyards; field house designed by D. H. Burnham & Co.; all extant.

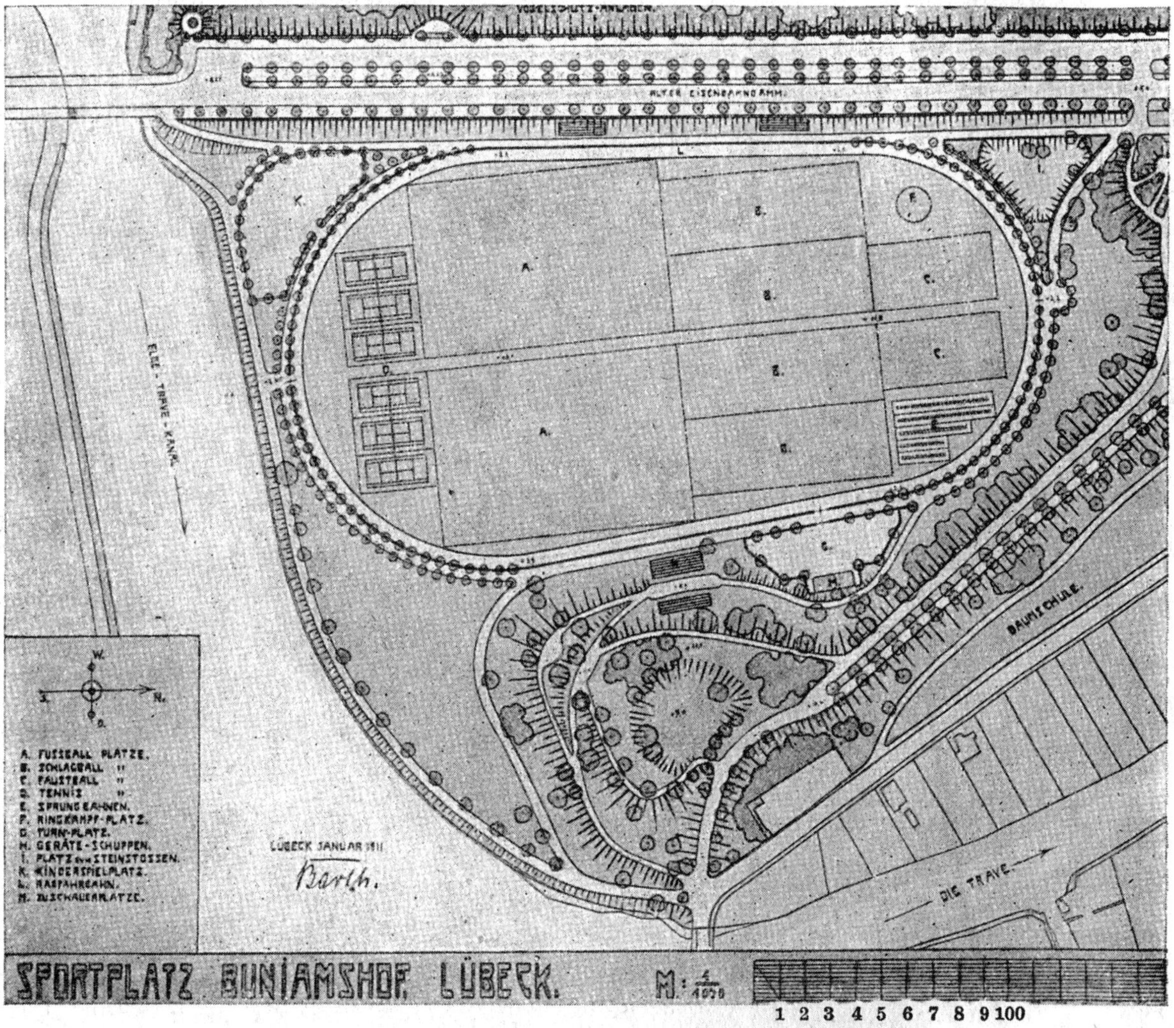

appendix 5

Outstanding example of a garden-like, rhythmic SPORT AREA, surrounded by a raised promenade. More than three hectares for actual play are divided into areas for football, baseball, volleyball, tennis, and for gymnastics, wrestling, and for the play of the little ones. Encircling this are lanes for bicycling, and competitive jumping and running. Naturally there are ample spaces for equipment and changing, as well as a protected area for spectators. (See p. 80.)

E. Barth, Charlottenburg

TRANSLATOR'S NOTE

Misprinted in the original as "H. Barth"; Erwin Barth began his career in Lübeck before moving to Berlin.

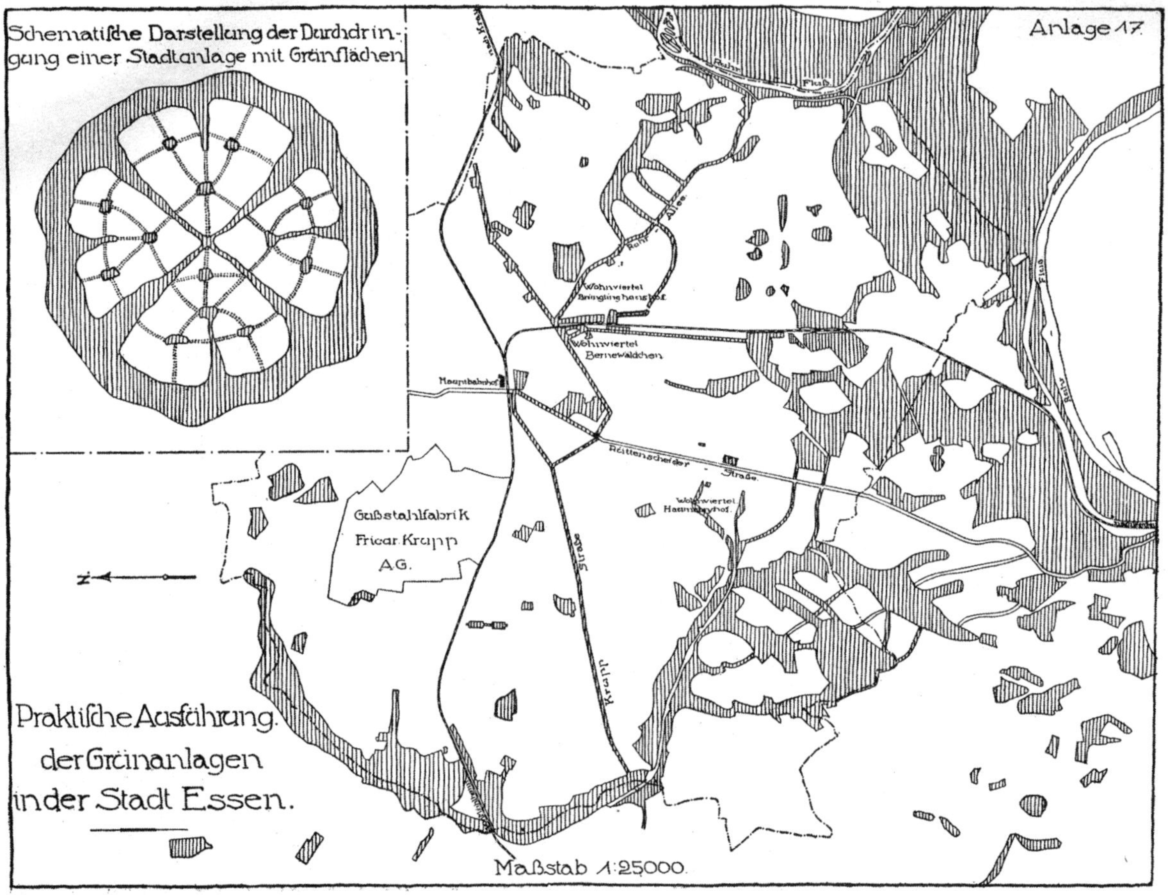

appendix 6

OPEN SPACES of the city of Essen, which was one of the first in Germany with a park policy mandating a planned system. (See p. 83.)

Councilor Schmidt, Essen

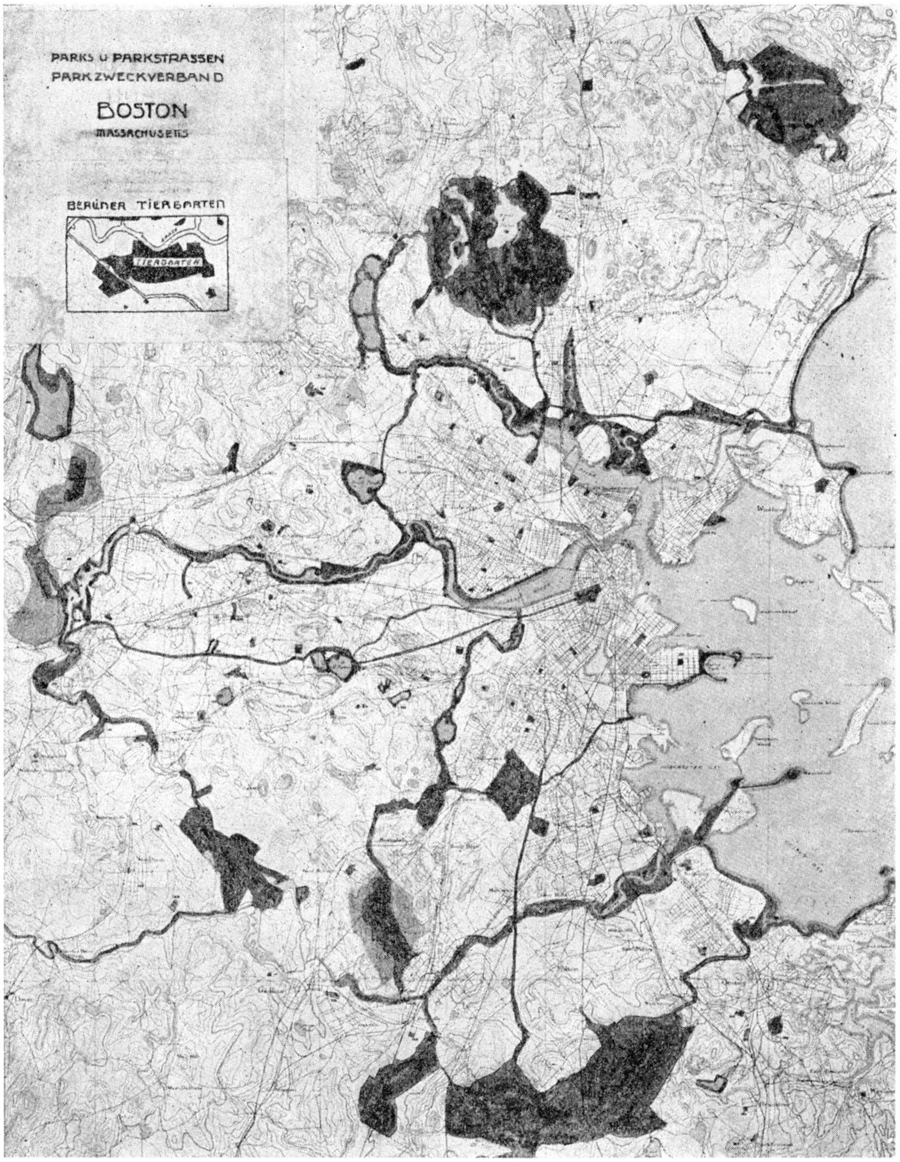

appendix 7

THE METROPOLITAN PARK COMMISSION of Greater Boston, which established its own open space system under the direction of the private landscape architect Charles Eliot, with an area of four hundred square kilometers and an overall cost of twenty million dollars.

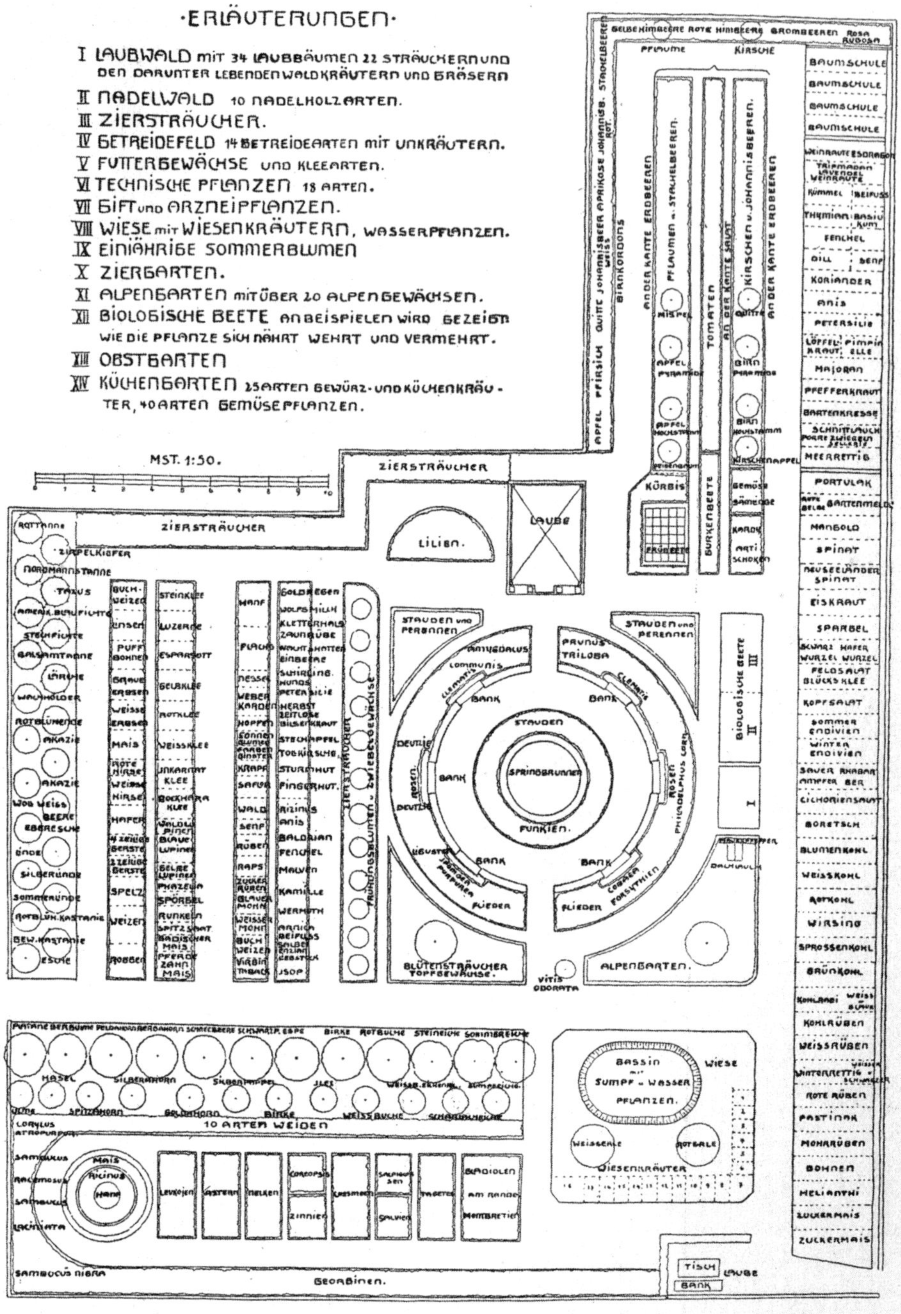

appendix 8

School garden of the School Garden Club of Pankow, Berlin, providing knowledge of useful and interesting plants in a simple, clear manner. (See p. 89.) This garden is the result of a private initiative.

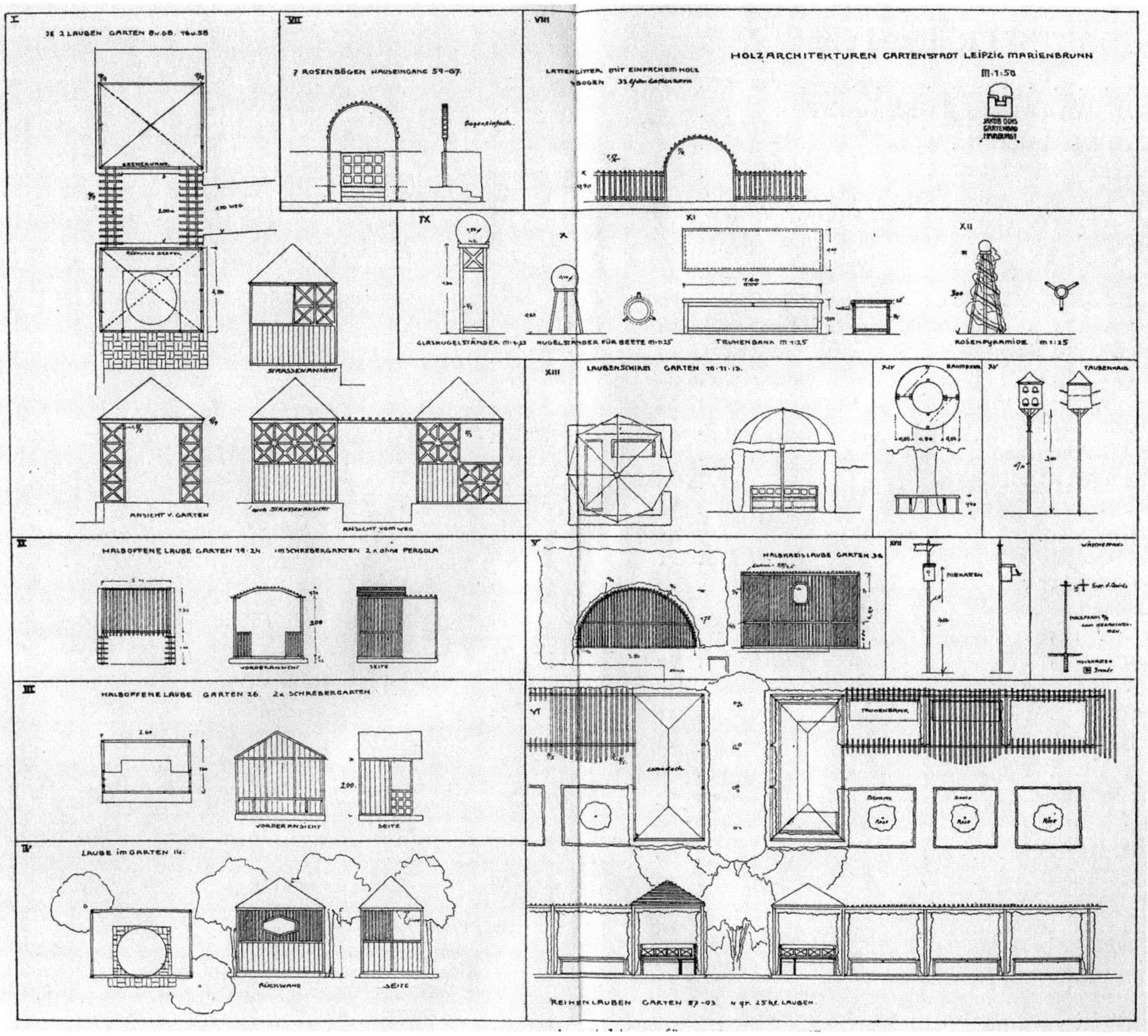

appendix 9

A collection of simple architectural structures for the small garden. Designed by the author for Marienbrunn Garden City in Leipzig.

PLANT NAMES IN *GARDEN CULTURE OF THE TWENTIETH CENTURY*

TRANSLATOR'S NOTES

The translator prepared these plant lists in collaboration with Dr. Clemens Alexander Wimmer, Berlin.

Plant names in the following tables are alphabetized according to the name(s) used in the English translation and subdivided according to major plant categories.

Migge sometimes refers to the same plant by different common names as well as by its Latin name. In most cases where plant names are given in Latin, these names are retained in the translation. German common names are translated into US English common names. Where Migge misspelled names in the original, the correct spelling is given in brackets. Many of the names used in the original publication could refer to various plants; in such cases, the most common plant names were chosen. Where both Latin and common names are used for the same plant, these are cross-listed in the tables below. In three cases, where Latin names are less common, the English common name has been added in brackets within the text and in the tables for clarification purposes.

	Plant name in English translation	Current scientific plant name	Plant name(s) in original German edition
Trees and shrubs	Alder	Alnus	Erle
	Amorpha [false indigo]	Amorpha fruticosa	Amorphen
	Apple	Malus	Apfel
	Arborvitae	Thuja occidentalis	Lebensbaum
	Ash, mountain	Sorbus aucuparia	Eberesche
	Ash, yellow mountain	Sorbus aucuparia "Dirkenii"	gelbe Eberesche
	Azalea	Rhododendron molle, & R. flavum	Azalee

	Plant name in English translation	Current scientific plant name	Plant name(s) in original German edition
Trees and shrubs (*continued*)	Beech	Fagus sylvatica	Buche; Rotbuche[1]
	Berberis	Berberis	Berberis
	Birch	Betula	Birke
	Blackberry	Rubus fruticosus	Brombeere
	Blackthorn	Prunus spinosa	Schlehe
	Bladder sennas	Colutea arborescens	Blasenstrauch
	Boxthorn	Lycium barbarum	Bocksdorn
	Boxwood (see Buxus)	Buxus	Buchsbaum
	Buckthorn	Rhamnus cathartica	Kreuzdorn
	Buckthorn, sea	Hippophae rhamnoides	Sanddorn
	Buxus	Buxus	Buxus
	Caragana	Caragana arborescens	Caragane
	Chamaecyparis squarrosa [Japanese false cypress]	Chamaecyparis squarrosa	Chamaecyparis squarrosa
	Cherry	Prunus avium	Kirsche
	Cherry, bird	Prunus padus	Traubenkirsche
	Cherry, white-variegated-leafed	?	weiß-buntblättrige Kirsche
	Chestnut	Aesculus hippocastanum	Kastanie[2]
	Chestnut, red flowering	Aesculus × carnea	Rot blühende Kastanie
	Cornus (see Dogwood)	Cornus	Kornus
	Cotoneaster	Cotoneaster	Quittenmispel[3]
	Currant (see Ribes)	Ribes	Johannisbeere
	Cypress, swamp	Taxodium distichum	Sumpfcypresse
	Daphne	Daphne	Daphne
	Deutzia	Deutzia	Deutzie
	Dogwood, yellow (see Cornus)	Cornus alba "Flaviramea"	gelbe Hartriegel
	Elderberry	Sambucus	Holunder
	Elm	Ulmus	Ulme
	Elm, Exeter	Ulmus glabra Exoniensis	Pyramiden-Ulme
	Fir	Abies	Tannenbaum
	Forsythia	Forsythia	Forsythie
	Gorse	Ulex europaeus	Ginster
	Hawthorn	Crataegus monogyna	Weißdorn
	Hazel	Corylus avellana	Hasel
	Hemlock, Canadian	Tsuga canadensis	Helmlockstanne

	Plant name in English translation	Current scientific plant name	Plant name(s) in original German edition
Trees and shrubs (*continued*)	Hornbeam	Carpinus betulus	Hainbuche
	Hydrangea	Hydrangea macrophylla	Hydrangie
	Ilex	Ilex aquifolium	Ilex
	Jasmine, winter	Jasminum nudiflorum	früher Jasmin
	Juniper	Juniperus communis	Wachholder
	Laburnum	Laburnum anagyroides	Goldregen
	Larch, Japanese	Larix kaempferi	Japanische Lärche
	Laurel (see Laurus)	Laurus nobilis	Loorbeer
	Laurus (see Laurel)	Laurus nobilis	Laurus
	Lilac	Syringa vulgaris	Flieder
	Linden	Tilia	Linde
	Linden, Crimean	Tilia × euchlora	Krimlinde
	Linden, large-leafed	Tilia platiphyllos	großblättrige Linde
	Linden, silver	Tilia tomentosa	Silberlinde
	Linden, small-leafed	Tilia parvifolia	kleinblättrige Linde
	Locust	Robinia pseudoacacia	Akazie
	Lonicera	Lonicera	Lonicere
	Magnolia	Magnolia	Magnolie
	Mahonia	Mahonia aquifolium	Mahonie
	Maple	Acer	Ahorn
	Maple, ash-leafed	Acer negundo "Argenteovariegatum"	weißbuntblättriger Ahorn
	Maple, globe Norway	Acer platanoides "Globosum"	Kugelahorn
	Maple, red	Acer platanoides "Schwedleri"	roter Ahorn
	Medlar	Mespilus germanica	Mispeln
	Oak	Quercus	Eiche
	Oak, pyramidal English	Quercus robur "Fastigiata"	Pyramideneiche
	Oak, scarlet	Quercus coccinea	Scharlacheiche
	Oleaster	Elaeagnus	Oelweiden
	Orange, mock	Philadelphus coronarius	Jasmin[4]
	Peach	Prunus persica	Pfirsich
	Pine	Pinus	Kiefer
	Pine, Mugo	Pinus mugo	Legföhren
	Pine, Pitch	Pinus rigida	Pitchpine
	Plane tree	Platanus × acerifolia	Platane

	Plant name in English translation	Current scientific plant name	Plant name(s) in original German edition
Trees and shrubs (*continued*)	Poplar	Populus	Pappel
	Poplar, pyramidal	Populus nigra "Italica"	Pyramidenpappel
	Poplar, silver	Populus alba	Bolleana-Pappel; Silberpappel
	Privet, common	Ligustrum vulgare	gemeiner Liguster
	Privet, oval-leafed	Ligustrum ovalifolium	halb immergrüner Liguster
	Quince	Cydonia oblonga	Quitte
	Quince, Japanese	Chaenomeles japonica	japanische Quitte
	Redwood, Sierra	Sequoiadendron giganteum	Wellingtonie
	Rhododendron	Rhododendron ponticum	Rhododendron
	Ribes (see Currant)	Ribes	Ribes
	Snowberry	Symphoricarpos albus	Schneebeere
	Spice bush	Calycanthus floridus	Gewürzstrauch
	Spirea	Spiraea	Spiräe
	Spruce	Picea abies	Fichte
	Spruce, dwarf Norway	Picea abies "Pumila"	Zwergfichte
	Spruce, serpentine	Picea abies "Virgata"	Schlangenfichte
	Sumac	Rhus typhina	Sumach
	Taxus (see Yew)	Taxus baccata	Taxus
	Tree of heaven	Ailanthus altissima	Götterbaum
	Tulip tree	Liriodendron tulipifera	Tulpenbaum
	Weigelia	Weigela	Weigelie
	Willow	Salix	Weide
	Willow, weeping golden	Salix alba "Tristis"	hängende Dotterweide
	Wingnut	Pterocarya fraxinifolia	Flügelnuß
	Yew (see Taxus)	Taxus baccata	Eibe
Flowers and herbaceous	Achillea	Achillea millefolium	Achillee
	Alyssum	Lobularia maritima	Alyssum
	Amaranth	Amaranthus caudatus	Fuchsschwanz
	Anemone	Anemone	Anemone
	Arabis	Arabis alpina	Arabis
	Asarum	Asarum europaeum	Asarum
	Aster	Aster	Aster
	Aster, China	Callistephus chinensis	Sommeraster
	Auricula	Primula auricula	Aurikel
	Baby's breath	Gypsophila paniculata	Schleierkraut

	Plant name in English translation	Current scientific plant name	Plant name(s) in original German edition
Flowers and herbaceous *(continued)*	Basil	Ocimum basilicum	Basilikum
	Bellflower	Campanula	Glockenblume
	Carnation	Dianthus caryophyllus	Nelke
	Centaurea	Centaurea montana	Centauree
	Chrysanthemum	Chrysanthemum × grandiflorum	Chrysantheme
	Cineraria	Senecio cineraria	Cinerarie
	Clover	Trifolium	Klee
	Coleus	Coleus blumei	Coleus
	Cornflower	Centaurea cyanus	Kornblume
	Crocus	Crocus	Krokus
	Dahlia	Dahlia	Dahlie
	Daisy	Bellis perennis	Gänseblümchen
	Dandelion	Taraxacum officinale	Löwenzahn
	Dimorphoteca	Dimorphotheca sinuata	Dimorphoteca
	Eschscholzia	Eschscholzia californica	Eschscholzie
	Flax, crimson	Linum grandiflorum	roter Lein
	Forget-me-not	Myosotis	Vergißmeinnicht
	Forget-me-not, water	Myosotis scorpioides	Sumpfvergißmeinnicht
	Foxglove	Digitalis purpurea	Fingerhut
	Foxglove, monstrous	Digitalis purpurea "Monstrosa"	monströser Fingerhut
	Fuchsia	Fuchsia	Fuchsie
	Gladiola	Gladiolus × hortulanus	Gladiole
	Godetia	Clarkia amoena	Godetie
	Gold-ball	Aurinia saxatalis var. plena	Goldball
	Goldenrod	Solidago canadensis	Goldraute
	Heliotrope	Heliotropium arborescens	Heliotrop
	Hollyhock	Alcea rosea	Malve; Stockrose
	Hyacinth	Hyacinthus orientalis	Hyazinthe
	Hypericum	Hypericum perforatum	Hypericee
	Impatiens	Impatiens balsamina	Balsamine
	Iris	Iris germanica	Iris
	Larkspur	Delphinium × elatium	Rittersporn
	Lavatera	Lavatera	Lavathere [Lavatere]
	Lavender	Lavandula angustifolia	Lavendel
	Lilac	Syringa vulgaris	Flieder

	Plant name in English translation	Current scientific plant name	Plant name(s) in original German edition
Flowers and herbaceous (*continued*)	Lily	Lilium	Lilie
	Lily, water (see Nymphaea)	Nymphaea alba	Seerose
	Lily of the valley	Convallaria majalis	Maiblume
	Liverwort	Hepatica nobilis	Leberblümchen
	Lobelia	Lobelia erinus	Lobelie
	Lupine	Lupinus polyphyllus	Lupine
	Lychnis	Lychnis coronaria	Lychnis
	Marigold	Tagetes patula	Studentenblume
	Mint	Mentha	Minze
	Morning glory	Ipomoea	Winde
	Narcissus	Narcissus	Narzisse
	Nasturtium	Tropaeolum majus	Kresse
	Nymphaea (see Lily, water)	Nymphaea	Nymphaea
	Orchid	Orchis	Orchidee
	Pansy	Viola tricolor	Stiefmütterchen
	Papaver nudicaule [Icelandic poppy]	Papaver nudicaule	Papaver nudicaule
	Peony	Paeonia officinalis	Pfingstrose
	Periwinkle	Vinca	Immergrün
	Petunia	Petunia × hybrida	Petunie
	Phacelia	Phacelia tanacetifolia	Phazelie
	Phlox	Phlox paniculata	Phlox
	Poppy	Papaver rhoeas	Mohn
	Poppy, Oriental	Papaver orientale	Orientalischer Mohn
	Portulaca	Portulaca grandiflora	Portulack [Portulak]
	Primrose	Primula vulgaris	Primel
	Pyrethrum	Tanacetum coccineum	Pyretrum [Pyrethrum]
	Reseda	Reseda odorata	Resede
	Rosemary	Rosmarinus officinalis	Roßmarien [Rosmarin]
	Sage	Salvia officinalis	Salvie
	Sainfoin	Onobrychis viciifolia	Esparsette
	Scabiosa	Scabiosa atropurpurea	Skabiose
	Scilla	Scilla sibirica	Scylla [Scilla]
	Sedum	Sedum	Sedum
	Snapdragon	Antirrhinum majus	Löwenmaul
	Snowdrop	Galanthus nivalis	Schneeglöckchen

	Plant name in English translation	Current scientific plant name	Plant name(s) in original German edition
Flowers and herbaceous (*continued*)	Stachys	Stachys byzantina	Stachis [Stachys]
	Stock	Matthiola incana	Levkoje
	Sunflower	Helianthus annuus	Sonnenblume
	Thyme	Thymus	Tymian [Thymian]
	Tickseed	Coreopsis	Schöngesicht
	Tobacco	Nicotiana alata	Tabak
	Tritoma	Kniphofia uvaria	Tritome
	Tuberose	Polianthes tuberosa	Tuberrose [Tuberose]
	Tulip	Tulipa gesneriana	Tulpe
	Veronica	Veronica	Veronike
	Violet	Viola	Veilchen
	Violet, horned	Viola cornuta	Gehörntes Veilchen
	Wallflower	Erysimum cheiri	Goldlack
	Zinnia	Zinnia peruviana	Zinnie
Roses (general)	Rose	Rosa	Rose
	Austrian copper rose	Rosa foetida	Kapuziner-Rose
	Cabbage rose	Rosa centifolia	Provinzrose
	Centifolia	Rosa centifolia	Centifolia
	Cinnamon rose	Rosa cinnamomea	Zimtrose
	Climbing rose	—	Schlingrose
	Dog rose	Rosa canina	Hundsrose
	Field rose	Rosa arvensis	Feldrose
	Moss rose	Rosa centifolia "Muscosa"	Moosrose
	Multiflora rose	Rosa multiflora	Rosa multiflora
	Niphetos	Rosa "Niphetos"	Niphitos
	Rugosa rose	Rosa rugosa	Rugose
	Sweetbriar rose	Rosa rubiginosa	schottische Zaunrose
	Tea rose	Rosa villosa	Teerose
	Villosa	Rosa villosa	Villosa
Rose cultivars	Ännchen Müller		
	Étoile de Lyon		
	Frau Geheimrat von Boch		
	Frau Karl Druschki		
	Gloire de Dijon		
	Gruß an Teplitz		

	Plant name in English translation	Current scientific plant name	Plant name(s) in original German edition
Rose cultivars (*continued*)	Herzogin Maria von Ratibor		
	Madame Jean Dupuy		
	Mme. Jules Gravereaux		
	Maréchal Niel		
	Souvenir de la Malmaison		
Vines	Actinidia	Actinidie	Actinidie
	Akebia, five-leaf	Akebia quinata	Akebia
	American bittersweet	Celastrus scandens	Baumwürger
	Aristolochia	Aristolochia macrophylla	Aristolochie
	Clematis	Clematis	Clematis
	Clematis, wild	Clematis vitalba	Wald-Clematis
	Cobaea	Cobaea scandens	Cobäe
	Grapevine, common	Vitis vinifera	edler Wein
	Grapevine, wild	Vitis	wilder Wein
	Honeysuckle	Lonicera caprifolium	Geißblatt; Jelängerjelieber
	Ivy	Hedera helix	Efeu
	Ivy, Boston	Parthenocissus tricuspidata "Veitchii"	Veitchis Wein; Vitis veitchi
	Maurandia	Maurandia	Maurandie
	Scarlet runner bean	Phaseolus coccineus	Feuerbohne
	Sweet pea	Lathyrus odoratus	Wicke; Ziererbse
	Vitis muralis[5]	Vitis muralis	Vitis muralis
	Wisteria	Wisteria	Glycine
Vegetables	Asparagus	Asparagus officinalis	Spargel
	Rutabaga	Brassica napobrassica	Kohlrübe
	Turnip	Brassica rapa subsp. rapa	Steckrübe, Rübe

The following sources were used in identifying the German plant names:
Alfred Lehmann, *Unsere Gartenzierpflanzen* (Zwickau: Förster and Borries, 1907).
Ludwig Späth, *Späth-Buch 1720–1930* (Berlin: L. Späth, 1930).

1 "Rotbuche," which translates as "European beech," is also used in the original edition; in the translation, only "beech" is used.

2 "Kastinie" or "chestnut" in the original edition actually refers to "horse chestnut"; here, only "chestnut" is used, in keeping with the original, although it is a different species.

3 C. simonsii is mentioned in one location; elsewhere, "Quittenmispel" is used, which could refer to other cotoneasters.

4 In Germany, "jasmine" is commonly used to refer to what is also called "false jasmine," which is the same species commonly referred to as "mock orange" in North America.

5 Vitis muralis is closely related to Parthenocissus quinquefolia (Virginia creeper) but it is a different species, and this name is still in use.

ex horto

DUMBARTON OAKS TEXTS IN GARDEN AND LANDSCAPE STUDIES

Dumbarton Oaks Research Library and Collection, Washington, D.C.

Ex horto is devoted to classic works on the philosophy, art, and techniques of landscape design. Augmented with contemporary scholarly commentary, the series offers historical texts from numerous languages and reintroduces valuable works long out of print. The volumes cover a broad geographical and temporal range, from ancient Chinese poetry to twentieth-century German treatises, and constitute a library of historical sources that have defined the core of the field. By making these works newly available, the series provides unprecedented access to the foundational literature of garden and landscape studies.

Further information on Garden and Landscape Studies publications can be found at www.doaks.org/publications.

Garden Culture of the Twentieth Century
Leberecht Migge, author; and David H. Haney, editor and translator

Travel Report: An Apprenticeship in the Earl of Derby's Kitchen Gardens and Greenhouses at Knowsley, England
Hans Jancke, author; Joachim Wolschke-Bulmahn, editor; and Mic Hale, translator